Engineering instrumentation and control

Engineering instrumentation and control

J. A. Haslam, M.Sc., B.Sc., C.Eng., M.I.E.E.

G. R. Summers, M.Sc., B.Sc., C.Eng., M.I.Mech.E., M.Inst.M.C.

D. Williams, M.Sc., Ph.D., C.Eng., M.I.E.E.

The three authors are senior members of the academic staff in the Faculty of Engineering Technology, Stockport College of Further and Higher Education.

Edward Arnold
A member of the Hodder Headline Group
LONDON MELBOURNE AUCKLAND

Edward Arnold is a division of Hodder Headline PLC
338 Euston Road, London NW1 3BH

First published in the United Kingdom 1981

14 13 12 11 10
99 98 97 96 95 94

Distributed in the USA by Routledge, Chapman and Hall, Inc.
29 West 35th Street, New York, NY 10001

British Library Cataloguing in Publication Data
Haslam, J A
Engineering intrumentation and control
1. Mensuration   2. Automatic control
I. Title   II. Summers, G R   III. Williams, D
620'.0044   T50

ISBN 0-7131-3431-3

Typeset in 10/11 Linotron Times by Keyset Composition
Printed and bound in the United Kingdom by
J W Arrowsmith Ltd, Bristol

Contents

Preface

Engineering Instrumentation and Control was written for students taking BTEC Higher National Programmes in Instrumentation and Control. It also provides useful background to some of the modules for Higher National programmes in Plant and Process Plant Engineering, and is suitable as an introduction to the subject for undergraduates studying Control Engineering.

The book aims to give the student a basic background knowledge of instrumentation and control, a familiarity with practical industrial measurement and control systems, and an understanding of specifications which will aid in the informed selection of system components.

The contents of the book are divided into three main areas: instrumentation science, instrumentation technology, and control. Instrumentation science involves the fundamental principles of instrumentation and measurement and is dealt with in chapters 1 to 5. The second area of instrumentation technology covered in chapters 6 to 12 involves the application of these principles to specific areas of measurement, with examples of complete measuring systems given at the end of each chapter. Chapters 13 to 15 deal with the different types of control system and the components used. Finally, appendices contain standard proofs for students with the mathematical background to undertake further work in this field.

J. A. Haslam
G. R. Summers
D. Williams

Acknowledgements

The authors wish to acknowledge the permission of the following manufacturers for the reproduction of illustrations and tables and for their assistance in the preparation of the book: Bell and Howell Ltd (figs 8.3 and 11.6(a)); Bruel and Kjaer (Table 5.2 and figs 11.6(b) and 11.12); Bryans Southern Instruments Ltd (Table 5.1, figs 5.6, 5.7, and 5.16, and information in section 5.6.4); Darenth Weighing Equipment Ltd (figs 9.4 and 9.13); Dowty Hydraulic Units Ltd (figs 14.14 and 14.15 and information in section 14.2.2); Keithley Instruments Ltd (fig. 5.15); Druck Ltd (fig. 10.8); Edwards High Vacuum Ltd (figs 10.13 and 10.14 and information in section 10.8); Gould Instruments Ltd (fig. 7.7 and information in section 7.4.4); Gulton Europe Ltd (Table 12.2 and figs 12.2(a) and (b), 12.8, and 12.13); Hottinger Baldwin Messtechnik GmbH (figs 9.9, 10.7, and 10.17 and information in section 10.10.3); Huggenberger Zurich (fig. 4.5); Kent Industrial Measurements Ltd (fig. 5.15); Kistler Instruments Ltd (figs 9.11 and 9.12); Maywood Instruments Ltd (figs 9.10(a) and (b)); Mullard Ltd (fig. 3.10); Penny and Giles Conductive Plastics Ltd (fig. 6.3 and information in section 6.4.1); Penny and Giles Data Recorders Ltd (fig. 5.14); RS Components Ltd (fig. 7.16); Strainstall Ltd (figs 9.10(c) and (d)); Taylor Instruments Ltd (fig. 14.3); Tektronix UK Ltd (fig. 5.19 and information in section 5.7.1); Vibrometer Ltd (figs 10.11, 10.16, and 11.7).

Table 12.1 and fig. 12.5 are based on BS 4937:parts 1 to 4:1973 and parts 5 to 7:1974 and the definitions in chapter 2 are based on BS 1523: part 1:1967 and BS 2643:1955 by permission of the British Standards Institution, 2 Park Street, London W1A 2BS, from whom copies of the complete standards may be obtained.

1 Introduction

1.1 Definition and needs for instrumentation

Although metrology can be precisely defined as 'the science of measurement', instrumentation cannot be so clearly defined. In general, however, instrumentation is the application of instruments for monitoring, sensing, and measurement. Its purpose may be

a) product testing and quality control;
b) monitoring in the interest of health, safety, or costing;
c) part of a control system; or
d) research and development.

Although mechanical instruments are still commonly used in engineering, there is an increasing usage of electronic equipment, for the following reasons:

i) the very rapid speed of response of electronic devices; and
ii) the ease with which electrical signals may be increased in amplitude and transmitted over long distances.

The key to successful instrumentation is undoubtedly accurate measurement, which can be achieved only if the needs and limitations of the instrumentation systems are understood. Measurement can be defined as a comparison with a standard; thus one should always be aware of the basic standard against which the comparison is being made. Calibration and periodic routine maintenance of the instruments are prerequisites of accurate measurement, without which one may too often waste time and effort recording meaningless readings and results.

Instrumentation is a multidiscipline subject which embraces physics, thermodynamics, mechanics, fluids, chemistry, and electrical principles and as such is an interesting and often stimulating area of engineering. Indeed, the engineer who understands and applies instrumentation needs to know a 'little bit about everything'.

1.2 Standards and calibration

The adoption of a fundamental set of standards and units provides a common basis upon which measurement can be made. If these fundamental or reference standards are natural ones – definable in atomic-process terms – they may be accurately reproduced anywhere in the world rather than being kept in one particular laboratory under special conditions.

The process of comparing an instrument with a known standard is referred to as 'calibration'. Calibration is normally performed either by varying one input quantity with all other parameters kept constant and observing the resulting output variations or, possibly, by marking or graduating an output scale as the primary quantity is varied through its full range. In practice a 'primary-standard' instrument is used to calibrate a secondary or working-standard instrument which in turn is used to calibrate the device in use. The accuracy and calibration of each device is therefore traceable back to the fundamental standard via the secondary and primary standards.

The National Physical Laboratory (NPL) is the United Kingdom's national standards laboratory, which maintains the national primary standards in strict accordance with internationally agreed recommendations. The NPL contributes to the development and improvement of these standards and to the investigation of the fundamental physical systems for application in measurement. In addition, the Laboratory provides a calibration service either directly, where high-precision or specialised equipment is required, or through approved laboratories of the British Calibration Service.

The British Standards Institution (BSI) is responsible for the preparation and promulgation of British Standards (BS) for all sectors of industry and trade, covering test methods, terms, definitions and symbols, performance and specifications, codes of practice, and other technical matters. BSI represents the United Kingdom in the International Organisation for Standardisation (ISO), and international agreement is therefore embodied in British Standards whenever possible.

1.3 The measuring system

Any measuring system can be broadly represented by the three-block diagram, shown in fig. 1.1. The blocks represent the following functional elements:

a) a transducer,
b) a signal conditioner, and
c) a recorder or indicator.

These may be further subdivided into several other blocks so that a more detailed block diagram may be produced representative of all the individual functions

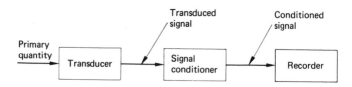

Fig. 1.1 Block diagram of a basic measurement system

The three-block-diagram approach should be used cautiously, since some systems and devices do deviate from such a generalisation. However, the basic approach does allow some categorisation of the various functions and elements which are available.

1.3.1 The transducer

The transducer element in fig. 1.1 is an energy converter which receives the physical quantity being measured (referred to as the *measurand*) and converts it into some other physical variable; e.g. flow to pressure, speed to voltage, strain to resistance. The transducer is undoubtedly the weakest link in the measuring chain, for the measured quantity is always modified by the presence of the transducer, making a perfect measurement theoretically impossible. This 'loading' effect can be minimised by the design and installation of the transducer, but it is always present to some degree.

1.3.2 The signal conditioner

The signal conditioner in fig. 1.1 rearranges the transduced signal into a form which can be readily recorded or monitored.

1.3.3 The recorder or display

The third element in the block diagram shown in fig. 1.1 is the recorder, display, or indicating device.

Many recorders have a transducing action at their input followed by some further signal conditioning. For example, an electromagnetic voltmeter – a recording device – transduces the input voltage into the displacement of a pointer moving over a calibrated scale. Even so, the voltmeter is usually represented by a single block as a recorder or monitoring device.

The Bourdon-tube pressure gauge (shown in fig. 10.6) may be represented by the block diagram shown in fig. 1.2. The Bourdon tube transduces the input pressure to a displacement at its closed end. The linkage system and gears condition the tip displacement into a movement of the indicating pointer across a graduated scale.

Fig. 1.2 Block diagram of a Bourdon-tube pressure gauge

Similarly, the three-block-diagram approach may be applied to a clinical thermometer: the temperature is 'transduced' into volumetric expansion of a bulb of mercury, a capillary tube 'conditions' the expanding mercury, and a scale permits measurement of the temperature in terms of length.

1.4 Loading effects

Since most measuring devices or instruments require energy to operate them, they absorb energy from the source. The presence of the measuring device thus changes the characteristics of the quantity being measured; for example, connecting a tachogenerator on to a rotating mechanism increases the friction and inertia of the system and so decreases the speed of rotation.

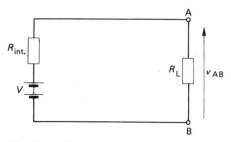

Fig. 1.3 Electrical loading effect

An example of this loading effect is shown in fig. 1.3, which shows a voltage source V having an internal resistance $R_{int.}$ and output terminals A and B. With terminals AB on open circuit, i.e. no load,

terminal voltage $v_{AB} = V$ volts

When a meter having a resistance R_L is connected across terminals AB, current I flows and

terminal voltage $v_{AB} = V - IR_{int.}$

but
$$I = \frac{V}{R_{int.} + R_L}$$

hence
$$v_{AB} = V \left(\frac{R_L}{R_L + R_{int.}} \right)$$

Note that, if R_L is much greater than $R_{int.}$,

$v_{AB} = V$

The conditions which must therefore be satisfied to minimise the loading effect are either

a) the source resistance must be small, or
b) the measuring-device resistance must be high.

Applying this to the general case, for accurate measurement the impedance of the measuring device should be much higher than the 'impedance' of the source, where impedance (opposition to flow) can be defined in general as

4

$$\text{impedance} = \frac{\text{force variable}}{\text{flow variable}}$$

Note that all measuring devices load the system to some extent. This problem can never be entirely eliminated and it should be taken into account.

1.5 General guide-lines

One should familiarise oneself with the needs and limitations of any instrumentation being used by referring to the manufacturers' handbooks, which usually detail the setting up, calibration, and operational procedures. 'When all else fails, read the instructions' is a maxim which unfortunately is too often applied by practising engineers. Time spent reading these handbooks is invaluable, since it helps to avoid inaccurate measurement or, even worse, damage to the instruments.

Where possible, measurement should always be begun with the instruments or the system on a low sensitivity or gain setting which can then be progressively increased until the desired sensitivity is reached. Overloading of the equipment may be avoided by using small loads, if possible, at the outset of the measurement. One should check and recheck. Never assume that the equipment is within the manufacturer's specification – test the calibration figures before and after the measurement.

Above all, cultivate good health-and-safety habits. The Health and Safety at Work Act (1974) imposes strict responsibilities on employer and employee alike to make them increasingly aware of hazards to health and safety. One should always be aware of possible hazards such as electric shock; chemical and gaseous poisoning; bodily damage; and burns caused by direct contact with or radiation from X-rays, ultra-violet light, or radioactive rays.

2 System performance

2.1 The ideal measuring system

The ideal measuring system is one where the output signal has a linear relationship with the measurand, where no errors are introduced by effects such as static friction, and where the output is a faithful reproduction of the input no matter how the input varies. This is, of course, a theoretical case and serves only as a comparison for actual results obtained from a measurement. Failure of a measuring system or instrument to match up to the perfect case is usually specified in terms of errors, where error is defined as the difference between the indicated value and the 'true value'.

The term 'true value' is taken here to mean the value obtained from an instrument or measuring system deemed by experts to be acceptably accurate for the purposes to which the results are being put. Thus, in calibrating a pressure gauge against a dead-weight tester, the readings from the latter would be taken as 'true values'.

It is convenient to examine measuring and control system performance in two ways:

a) when steady or constant input signals are applied, comparison of the steady output with the ideal case gives the *static performance* of the system;

b) when changing input signals are applied, comparison with the ideal case gives the *dynamic performance* of the system.

2.2 Static performance

2.2.1 Sensitivity

Static sensitivity is defined as the ratio of the change in output to the corresponding change in input under static or steady-state conditions.

$$\text{Static sensitivity } K = \frac{\Delta\theta_o}{\Delta\theta_i} \qquad 2.1$$

where $\Delta\theta_o$ = the change in output

and $\Delta\theta_i$ = the corresponding change in input

Sensitivity may have a wide variety of units, depending on the instrument or measuring system being considered. The platinum resistance thermometer, for example, gives a change of resistance with increase of temperature and therefore its sensitivity would have units of ohms/°C.

Figure 2.1(a) shows a linear relationship between output and input, and sensitivity therefore equals the slope of the calibration graph. In the case of the non-linear input/output relationship shown in fig. 2.1(b), the sensitivity will vary according to the value of the output.

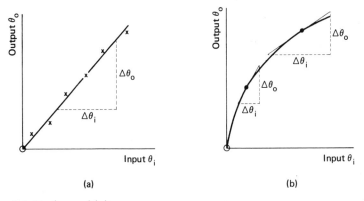

(a)   (b)

Fig. 2.1 Static sensitivity

Manufacturers of recording and display equipment tend to quote values of sensitivity which are the inverse of those given by the above definition; for example, an oscilloscope sensitivity would be quoted in V/cm rather than cm/V which would be expected from the definition.

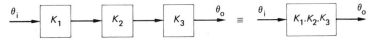

Fig. 2.2 Overall system sensitivity

If elements of a system having static sensitivities of $K_1, K_2, K_3, \ldots$, etc. are connected in series or cascade as illustrated in fig. 2.2, then the over-all system sensitivity K is given by

$$K = K_1 \times K_2 \times K_3 \times \ldots \qquad\qquad 2.2$$

provided that there is no alteration in the values of $K_1, K_2, K_3, \ldots$, etc. due to loading effects.

It is common to find system elements that have inputs and outputs of the same form – for example the voltage amplifier – and in this case the term *gain*, or more specifically voltage gain, is used rather than sensitivity. In mechanical systems using lever systems it is more usual to use the term *magnification* to describe the increase in displacement. The terms sensitivity, gain, and magnification all mean the same.

7

Example 2.1   A measuring system consists of a transducer, an amplifier, and a recorder, with individual sensitivities as follows:

transducer sensitivity   0.2 mV/°C
amplifier gain   2.0 V/mV
recorder sensitivity   5.0 mm/V

Determine the overall system sensitivity.

Using equation 2.2,

$$K = K_1 \times K_2 \times K_3$$

$$= 0.2\,mV/°C \times 2.0\,V/mV \times 5.0\,mm/V$$

$$= 2.0\ mm/°C$$

2.2.2 Accuracy and precision

Since all measurement involves error, the question 'Is the system accurate?' is meaningless, since it can always be answered in the negative. What is more important is the answer to the question 'How accurate is the system?'

Accuracy is normally stated in terms of the errors introduced, where

$$\text{percentage error} = \frac{\text{indicated value} - \text{true value}}{\text{true value}} \times 100\%$$

However, it is common practice to express the error as a percentage of the measuring range of the equipment:

$$\text{percentage error} = \frac{\text{indicated value} - \text{true value}}{\text{maximum scale value}} \times 100\% \qquad 2.3$$

For example, if a 0 to 1 bar pressure gauge is accurate to within ±5% of full-scale deflection, then the maximum error will be ±0.05 bar. Note that if the gauge is used at the lower end of its range then the ±0.05 bar error will result in a larger percentage error from the 'true value'.

Example 2.2   A 0 to 10 bar pressure gauge was found to have an error of ±0.15 bar when calibrated by the manufacturer. Calculate (a) the percentage error of the gauge and (b) the possible error as a percentage of the indicated value when a reading of 2.0 bars was obtained in a test.

a) Using equation 2.3,

$$\text{percentage error} = \frac{0.15\ bar}{10\ bar} \times 100\%$$

$$= \pm 1.5\%$$

8

b)   Possible error $= \pm 0.15$ bar

$\therefore$ error at 2.0 bars $= \pm \dfrac{0.15 \text{ bar}}{2.0 \text{ bar}} \times 100\%$

$\qquad\qquad\qquad = 7.5\%$

The gauge is therefore more unreliable at the lower end of its range, and an alternative gauge with a more suitable range should be used.

'Precision' is a term that is sometimes confused with accuracy, but a precise measurement may not be an accurate measurement. If the measuring device is subjected to the same input on a number of occasions and the indicated results lie closely together, then the instrument is said to be of high precision. The term used to specify the closeness of results is the *reproducibility* of the instrument, and this will be dealt with in section 2.2.4.

If a good-quality voltmeter is used to measure a constant voltage on a number of different occasions and (within the limits of accuracy of the instrument) all the readings are the same, then precise readings are said to have been obtained. Suppose, however, that when putting the voltmeter away after the test it is noticed that the pointer is offset and not reading zero. All the readings obtained would be precise but not accurate. Figure 2.3 illustrates diagrammatically the difference between the two terms.

X = result   Centre circle represents true value

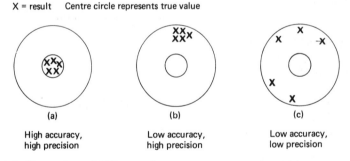

(a)	(b)	(c)
High accuracy, high precision	Low accuracy, high precision	Low accuracy, low precision

Fig. 2.3 Illustration of difference between accuracy and precision

2.2.3 Possible and probable errors
Consider a measurement that involves the use of three devices with maximum possible errors of $\pm a\%$, $\pm b\%$, and $\pm c\%$ respectively. It is unlikely that all three devices will have their maximum errors at the same time; therefore a more practical way of expressing the overall system error is to take the square root of the sum of the squares of the individual errors,

9

i.e. $\quad$ root-sum-square error of overall system $\quad = \pm \sqrt{(a^2 + b^2 + c^2)}$ $\qquad$ 2.4

Example 2.3 For a general measuring system where the errors in the transducer, signal conditioner, and recorder are $\pm2\%$, $\pm3\%$, and $\pm4\%$ respectively, calculate the maximum possible system error and the probable or root-sum-square error.

Maximum possible error $= \pm(2 + 3 + 4)\% = \pm9\%$

Using equation 2.4,

$$\text{root-sum-square error} = \pm \sqrt{(2^2 + 3^2 + 4^2)}\%$$
$$= \pm \sqrt{29}\%$$
$$= \pm5.4\%$$

Thus the error is possibly as large as $\pm9\%$ but probably not larger than $\pm5.4\%$.

2.2.4 Other static-performance terms

Reproducibility A general term applied to the ability of a measuring system or instrument to display the same reading for a given input applied on a number of occasions.

Repeatability The reproducibility when a constant input is applied repeatedly at short intervals of time under fixed conditions of use.

Stability The reproducibility when a constant input is applied over long periods of time compared with the time of taking a reading, under fixed conditions of use.

Constancy The reproducibility when a constant input is presented continuously and the conditions of test are allowed to vary within specified limits, due to some external effect such as a temperature variation.

Range The total range of values which an instrument or measuring system is capable of measuring.

Span The range of input signals corresponding to the designed working range of the output signal.

Tolerance The maximum error.

Linearity The maximum deviation from a linear relationship between input and output – i.e. from a constant sensitivity – expressed as a percentage of full scale.

Resolution The smallest change of input to an instrument which can be detected with certainty, expressed as a percentage of full scale.

Dead-band The largest change of input to which the system does not respond due to friction or backlash effects, expressed as a percentage of full scale.

Hysteresis The maximum difference between readings for the same input when approached from opposite directions – i.e. when increasing and decreasing the input – expressed as a percentage of full scale.

2.3 Dynamic performance

Many industrial processes require the measurement of parameters which remain constant or change very slowly, for example a constant pressure or temperature in a chemical process. In cases like this, the static performance of the measuring system is of prime concern. However, with the increase of automation, a greater emphasis is being placed on whether a device can respond adequately to changing signals. If a transducer responds sluggishly to a sudden change of input parameter, then automatic control of that parameter may become difficult if not impossible. As a further example, consider a vibration-measuring system, where the parameter is by its very nature a changing quantity. If the system could not respond to the frequencies of the vibration then the results would be totally useless.

The dynamic performance of both measuring and control systems is extremely important and is specified by responses to certain standard test inputs. These are

a) the *step input*, which takes the form of an abrupt change from one steady value to another. This indicates how well the system can cope with the change and results in the *transient response*.
b) the *ramp input*, which varies linearly with time and gives the *ramp response*, indicating the steady-state error in following the input.
c) the *sine-wave input*, which gives the *frequency response* or *harmonic response* of the system. This shows how the system can respond to inputs of a cyclic nature as the frequency f Hz or ω rad/s (where $\omega = 2\pi f$) varies.

These three inputs are illustrated in fig. 2.4 and will be encountered again in chapter 15. However, in this chapter only the transient and frequency responses will be considered, since these will be needed to interpret measuring-system specifications.

All systems will to some extent fail to follow exactly a changing input, and a measure of how well a system will respond is indicated by its dynamic specifications. These are expressed as step or transient parameters or frequency-response parameters, depending on the type of input applied. Many systems, although different in nature, produce iden-

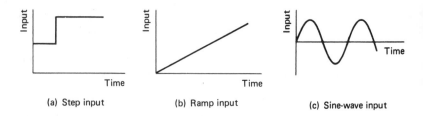

Fig. 2.4 Different forms of dynamic test inputs

tical forms of response, and this is due to the fact that the system dynamics are similar – i.e. the dynamic or differential equations are of the same form.

2.3.1 Zero-order systems

The ideal measuring system mentioned in section 2.1 is one whose output is proportional to the input no matter how the input varies, i.e. the mathematical equation relating them is of the form

$$\theta_o = K\theta_i \qquad 2.5$$

where K is the sensitivity of the system.

This is the equation of a zero-order system, since there are no differential coefficients present. Alternatively, equation 2.5 can be expressed in terms of the ratio θ_o/θ_i to give

$$\frac{\theta_o}{\theta_i} = K \qquad 2.6$$

and this can be represented by the block diagram shown in fig. 2.5.

Fig. 2.5 Block-diagram representation of a zero-order system

In practice the measuring system which approaches the ideal zero-order system is the potentiometer, which gives an output voltage proportional to the displacement of the wiper.

2.3.2 First-order systems

A first-order system is one whose input/output dynamics are represented by a first-order differential equation of the form

$$a \frac{d\theta_o}{dt} + b\theta_o = c\theta_i \qquad 2.7$$

12

where θ_o is the output variable;
        θ_i is the input variable;

and $a, b,$ and c are constants.

This equation can be rewritten to obtain a unity coefficient of θ_o, giving

$$\frac{a}{b} \cdot \frac{d\theta_o}{dt} + \theta_o = \frac{c}{b} \cdot \theta_i$$

and this can be expressed in a standard form as

$$\tau \frac{d\theta_o}{dt} + \theta_o = K\theta_i \qquad\qquad 2.8$$

where τ is the time constant in seconds

and K is the static sensitivity, with the units of the ratio θ_o/θ_i.

Comparison of equations 2.7 and 2.8 shows that

$$\tau = \frac{a}{b} \quad \text{and} \quad K = \frac{c}{b}$$

Equation 2.8 can be written in terms of the D operator, where

$$D \equiv \frac{d}{dt} \quad \text{and} \quad D^2 \equiv \frac{d^2}{dt^2} \quad \text{etc.}$$

hence $\tau D \theta_o + \theta_o = K\theta_i$

i.e. $(\tau D + 1)\theta_o = K\theta_i$

$$\therefore \qquad \frac{\theta_o}{\theta_i} = \frac{K}{1 + \tau D} \qquad\qquad 2.9$$

The ratio θ_o/θ_i expressed in terms of the D operator is known as the *transfer operator* of the system.

For the first-order system, equation 2.9 represents the standard form of the transfer operator and can be represented by the block diagram in fig. 2.6. Note that equations 2.7, 2.8, and 2.9 are all alternative ways of expressing the same differential equation.

θ_i  θ_o

Fig. 2.6 Block-diagram representation of a first-order system

Examples of first-order systems include

a) the mercury-in-glass thermometer, where the heat conduction through the glass bulb to the mercury is described by a first-order differential equation;

13

b) the build-up of air pressure in a restrictor/bellows system;
c) a series resistance–capacitance network.

Example 2.4 The differential equation describing a mercury-in-glass thermometer is

$$4\frac{d\theta_o}{dt} + 2\theta_o = 2 \times 10^{-3}\,\theta_i$$

where θ_o is the height of the mercury column in metres and θ_i is the input temperature in °C. Determine the time constant and the static sensitivity of the thermometer.

For the standard form shown in equation 2.8, the θ_o term must have unity coefficient and therefore dividing all terms by 2 gives

$$2\frac{d\theta_o}{dt} + \theta_o = 10^{-3}\,\theta_i$$

Comparing this with equation 2.8,

i.e. $\tau\dfrac{d\theta_o}{dt} + \theta_o = K\theta_i$

it can be seen that

$$\tau = 2\,s$$

and $K = 10^{-3}\,m/°C$ or $1\,mm/°C$

The standard form indicated in equations 2.8 and 2.9 is very convenient, because a first-order system always produces a standard response to either a step or a sine-wave input.

a) Step response
Figure 2.7(a) shows the exponential rise to the final value which is characteristic of the first-order system. The dynamic error is the difference between the ideal and actual responses, and comparison of the two shows that this error decreases with time. The step response is shown in more detail in fig. 2.7(b), and from this the definition of time constant (τ) is obtained:

> time constant is the time taken to reach the final value if the initial rate had been maintained, or the time taken to reach 63.2% of the step change.

It is important to note the initial slope at time $t = 0$, as this distinguishes it from a second-order step response, which has zero initial slope.

14

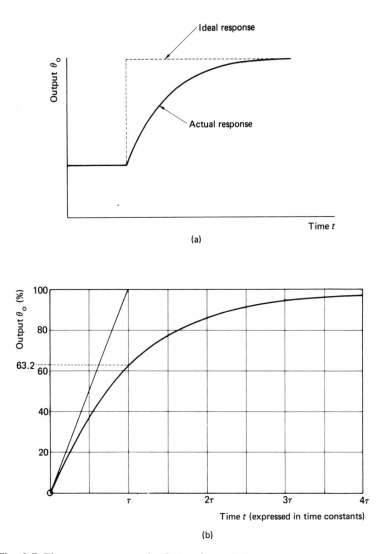

Fig. 2.7 The step response of a first-order system

b) Frequency response

This response is obtained by applying sine waves of a known amplitude at the input and examining how the output responds as the frequency of the sine wave is varied. Figure 2.8 illustrates the inability of the system to follow the input faithfully, and it can be seen that the output lags behind the input. As the frequency is increased, the output falls further behind and – perhaps more importantly – decreases in amplitude.

15

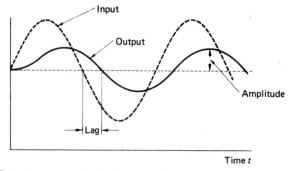

Fig. 2.8 Output response to a sine-wave input

The ratio of output amplitude to input amplitude is called the *amplitude ratio* and should equal a constant irrespective of the input frequency, i.e. $\theta_o = K\theta_i$ at all times. Figure 2.9 illustrates the standard frequency-response curve for a first-order system, showing the variation of amplitude ratio with frequency. It is valid for $K = 1$ or if the amplitude ratio is regarded as the ratio of actual output amplitude to the ideal output amplitude. The frequency axis is in generalised $\omega\tau$ form and enables the curve to be used for different values of time constant. To obtain the ω values in rad/s for a particular time constant τ, the horizontal-axis units would have to be multiplied by the factor $1/\tau$.

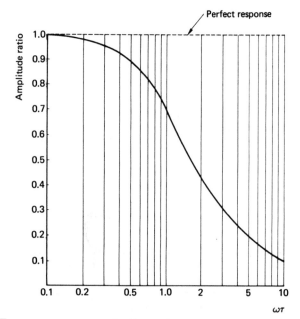

Fig. 2.9 Frequency response of a first-order system

Example 2.5 If a first-order measuring system has a time constant of 0.01s, determine the approximate range of input signal frequencies that the system could follow to within 10%.

From fig. 2.9 the amplitude ratio is greater than 0.9, i.e. there is less than 10% error, up to $\omega\tau \simeq 0.5$.

$\therefore$ $\omega\tau = 0.5$ defines the upper frequency limit

i.e. $\omega = \dfrac{0.5}{0.01s} = 50\,\text{rad/s}$

therefore the required frequency range is from 0 to $(50/2\pi)\,\text{Hz}$, i.e. from 0 to 8 Hz.

Example 2.6 A first-order system has a time constant of 6ms. Determine the frequency corresponding to the condition $\omega\tau = 1$ and calculate the approximate percentage error at this frequency.

$\omega\tau = 1$

$\therefore$ $\omega = \dfrac{1}{\tau}$

$\quad = \dfrac{1}{6 \times 10^{-3}\text{s}} = 166.7\,\text{rad/s}$

$\therefore$ frequency $f = \dfrac{166.7}{2\pi} = 26.5\,\text{Hz}$

From fig. 2.9, the amplitude ratio $\simeq 0.7$ at $\omega\tau = 1$

$\therefore$ % error = 30%

2.3.3 Second-order systems

A second-order system is one whose input/output relationship is described by

$$a\,\frac{d^2\theta_o}{dt^2} + b\,\frac{d\theta_o}{dt} + c\theta_o = e\theta_i \qquad\qquad 2.10$$

where a, b, c and e are constants.

This can be rearranged to give a unity coefficient of the $d^2\theta_o/dt^2$ term,

i.e. $\dfrac{d^2\theta_o}{dt^2} + \dfrac{b}{a}\cdot\dfrac{d\theta_o}{dt} + \dfrac{c}{a}\cdot\theta_o = \dfrac{e}{a}\cdot\theta_i$

and this can be written in a standard form as

$$\frac{d^2\theta_o}{dt^2} + 2\zeta\omega_n\cdot\frac{d\theta_o}{dt} + \omega_n^2\cdot\theta_o = K'\theta_i \qquad\qquad 2.11$$

17

where ω_n is the undamped natural frequency in rad/s,

ζ ('zeta') is the damping ratio,

and K' is a constant which would equal ω_n^2 if θ_o and θ_i were equal under static conditions.

In terms of the D operator, equation 2.11 becomes

$$(D^2 + 2\zeta\omega_n D + \omega_n^2)\theta_o = K'\theta_i$$

giving the transfer operator

$$\frac{\theta_o}{\theta_i} = \frac{K'}{D^2 + 2\zeta\omega_n D + \omega_n^2} \qquad 2.12$$

Alternatively, dividing through by ω_n^2 gives

$$\frac{\theta_o}{\theta_i} = \frac{K}{(1/\omega_n^2)D^2 + (2\zeta/\omega_n)D + 1} \qquad 2.13$$

where K is the static sensitivity.

This form of the transfer operator can be represented by the block diagram shown in fig. 2.10.

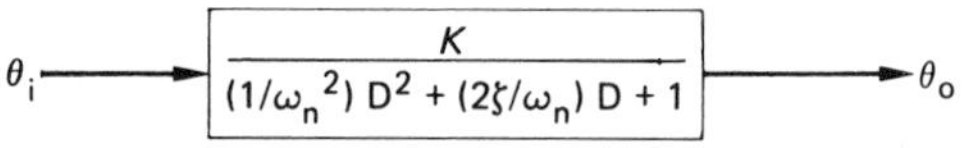

$$\theta_i \longrightarrow \boxed{\frac{K}{(1/\omega_n^2)\,D^2 + (2\zeta/\omega_n)\,D + 1}} \longrightarrow \theta_o$$

Fig. 2.10 Block-diagram representation of a second-order system

ω_n is a measure of the speed of response of the second-order system – higher values of ω_n would mean that the system would respond more rapidly to sudden changes.

ζ is a measure of the damping present in a system and is equal to the ratio of actual damping to critical damping. Its value determines the forms of the step and frequency responses, as follows.

a) *When $\zeta < 1$* The system is here said to be underdamped and results in oscillations occurring in the step response and (for values of $\zeta < 0.707$) resonance effects in the frequency response. 'Resonance' here means an output signal greater in amplitude than the ideal output.

b) *When $\zeta = 1$* This is the critically damped condition; i.e. no oscillations or overshoots appear in the step response and there is no resonance in the frequency response. This is the point of change-over from an underdamped condition to an overdamped condition.

c) *When $\zeta > 1$* Here the system is overdamped and responds in a sluggish manner, again with no overshoot in the step response and no resonance in the frequency response.

The most common example of a second-order system is a mass–spring system with damping. A large number of devices and mechanisms are of this type, e.g. the u.v. galvanometer, the piezo-electric transducer, and the pen-control system on X–Y plotters.

a) Step response
The step response for second-order systems does not have one unique form but may have one of an infinite number depending on the value of ζ.

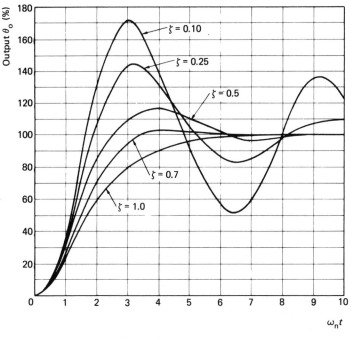

Fig. 2.11 Step responses of a second-order system

Figure 2.11 shows the step responses for a number of ζ values, and these illustrate the fact that as the damping in the system reduces so the overshoot and oscillation increase. A generalised time scale in terms of $\omega_n t$ is used to enable the same curves to apply to very fast systems with high ω_n values and to slow systems with low values of ω_n. A useful curve showing the relationship between the first overshoot and the damping ratio is shown in fig. 2.12, and this can be used to estimate ζ from an underdamped step response. The term 'percentage overshoot' is used to express the magnitude of the first overshoot as a percentage of the final or steady-state value.

Example 2.7 A mass–spring–damper system has a first overshoot of approximately 40% of its final value when subjected to a step input force.

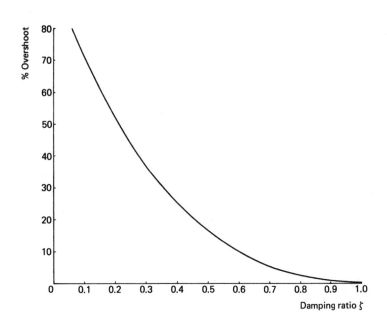

Fig. 2.12 Graph of % overshoot against damping ratio ζ

Estimate the values of the damping ratio ζ and ω_n if the time taken to reach the first overshoot is 0.8 s from the application of the step.

From fig. 2.12, $\zeta = 0.28$ results in a 40% overshoot. If an approximation of $\zeta = 0.25$ is made then fig. 2.11 can be used to determine ω_n:

$\omega_n t = 3.2$ units at the first overshoot

$$\therefore \quad \omega_n = \frac{3.2}{0.8\,\text{s}} = 4\,\text{rad/s}$$

The technique outlined in example 2.7 provides a reasonably quick and convenient method of estimating the value of ζ from the percentage overshoot of the step response, provided the standard curves are available.

As far as the best step response is concerned, it can be seen that lower ζ values exhibit a faster response but increased overshoot, while higher values result in very sluggish responses but no overshoot. The optimum condition is therefore a compromise between an acceptable speed of response and amount of overshoot, and a value of $\zeta \approx 0.7$ is usually stated for most second-order measurement systems.

b) Frequency response

Figure 2.13 illustrates a typical set of frequency-response curves for a second-order system. Examination shows that the amplitude-ratio axis is correct for $K = 1$, and a generalised-frequency axis is plotted to allow use of the curves for any second-order system. Resonance effects are observed for lightly damped cases as the frequency of the input approaches the natural frequency of the system. This abnormally high output response occurs only for damping ratios of less than 0.7. As the input frequency is increased well beyond the natural frequency, then the amplitude ratio falls as the system fails to respond to the higher rates of change.

The ideal frequency response for either a measuring system or an automatic-control system would be one which had an amplitude ratio of unity for all frequencies. The nearest response to this is for a ζ value of between 0.6 and 0.7, which has a constant amplitude ratio within ±3% of

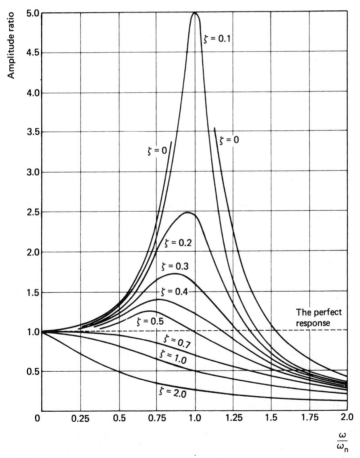

Fig. 2.13 Frequency response of a second-order system

unity for a range of frequencies up to 60% of the undamped natural frequency f_n.

In both the step and the frequency responses the optimum value of ζ lies between 0.6 and 0.7, with 0.64 often quoted, and manufacturers of measuring systems must ensure that the correct amount of damping is present. The word 'optimum' is used here to indicate the best response to step and sine-wave signals, but there are devices which are specifically designed with overdamping so as not to follow varying signals but to give average values. In these cases a higher value of ζ would be optimum.

Example 2.8 In a frequency-response test on a second-order system, resonance occurred at a frequency of 216 Hz, giving a maximum amplitude ratio of 1.36. Estimate the values of ζ and ω_n for the system.

Examination of fig. 2.13 indicates that $\zeta = 0.4$ gives an amplitude ratio of 1.4,

$$\therefore \quad \zeta \approx 0.4$$

This resonance occurs at $\omega/\omega_n = 0.8$

But $\omega = 2\pi f$

$\qquad = 2\pi \times 216 \, \text{Hz}$

$\qquad = 1357 \, \text{rad/s}$

$$\therefore \quad \omega_n = \frac{\omega}{0.8}$$

$$= \frac{1357 \, \text{rad/s}}{0.8} = 1696 \, \text{rad/s}$$

Note: Resonance occurs at frequencies which are lower than the undamped natural frequency except in the one case where the damping is zero.

2.3.4 Step-response specification

Three terms are used to specify a system's step response – namely response time, rise time, and settling time – and these are defined as follows.

Response time $(t_{res.})$ The time taken for the system output to rise from 0% to the first cross-over point of 100% of the final steady-state value. Applicable only to underdamped systems.

Rise time (t_r) The time taken for the system output to rise from 10% to 90% of its final steady-state value.

Settling time (t_s) The time taken for the system output to reach and *remain within* a certain percentage tolerance band of the final steady-state value. Typical values would be 2% and 5% settling times.

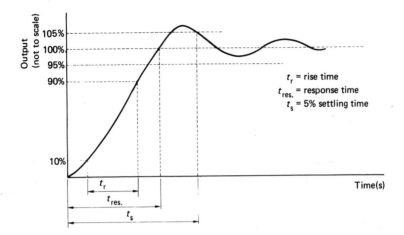

Fig. 2.14 Step response illustrating response, rise, and settling times

These parameters are illustrated in fig. 2.14, which shows a step response containing oscillations. In the figure, t_s refers to the 5% settling time, which is the time taken to reach and remain within the tolerance band of 95% to 105% of the final value.

2.3.5 Frequency-response specification
There are numerous ways of specifying the frequency responses of measuring systems, instruments, and control systems and this can lead to some confusion. A distinction can be made between devices which are designed to work over a range or band of frequencies – called a.c. devices – and those which can respond to d.c., i.e. down to zero frequency, called d.c. devices.

a) A.C. devices
For devices of this type, the gain or amplitude ratio is usually constant over a given frequency range, but at low or high frequencies the gain falls off, as illustrated in fig. 2.15. The term used to specify the frequency range is *bandwidth*, which is equal to $(f_2 - f_1)$ Hz.

> Bandwidth is the range of frequencies between which the gain or amplitude ratio is constant to within −3dB (this corresponds to a 30% reduction in gain).

Thus an a.c. amplifier on an oscilloscope will have a typical specification of a bandwidth of 8Hz to 10MHz, which means that at 8Hz and 10MHz the trace size for a constant-amplitude input wave will be 70% of that for a mid frequency, say 1kHz. Therefore at these frequencies the values read from the oscilloscope screen will be 30% lower than the actual values.

23

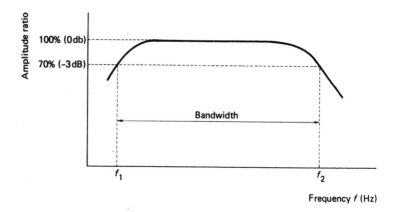

Fig. 2.15 Typical frequency response of an a.c. device

b) D.C. devices
In this context, d.c. device does not mean a device which will respond only to steady or d.c. signals but one which will respond to both a.c. and d.c. signals. A typical frequency response for devices of this type is shown in fig. 2.16 and might apply, for example, to a pen recorder. Here the operating range may be expressed as the upper frequency at which the gain or amplitude ratio falls outside a certain tolerance band. A value of ±3% is illustrated in fig. 2.16, but other common values are ±5%, −3dB as mentioned in (a), or ±1dB which is equivalent to ±10% (see appendix C).

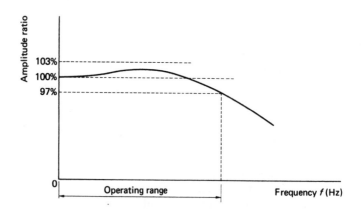

Fig. 2.16 Typical frequency response of a d.c. device

24

Exercises on chapter 2

1 A temperature-measuring system incorporates a platinum resistance thermometer, a Wheatstone bridge, a voltage amplifier, and a pen recorder. The individual sensitivities are as follows:

Transducer $0.35\,\text{ohm}/°\text{C}$
Wheatstone bridge $0.01\,\text{V/ohm}$
Amplifier gain $100\,\text{V/V}$
Pen recorder $0.1\,\text{cm/V}$

Determine (a) the overall system sensitivity from equation 2.2 and (b) the temperature change corresponding to a recorder pen movement of 4 cm.   [$0.035\,\text{cm}/°\text{C}$; $114.3°\text{C}$]

2 A 0–100°C thermometer is found to have a constant error of $+0.2°\text{C}$. Calculate the percentage error at readings of (a) 10°C, (b) 50°C, and (c) 100°C.   [$+2.0\%$; $+0.4\%$; $+0.2\%$]

3 A vibration-measuring system involves the use of a piezo-electric transducer, a charge amplifier, and a u.v. recorder. If the maximum errors are $\pm0.5\%$, $\pm1.0\%$, and $\pm1.5\%$ respectively, calculate the maximum possible system error and the probable or root-sum-square error.   [$\pm3.0\%$; $\pm1.87\%$]

4 Define the following terms applied to measuring systems: (a) reproducibility, (b) repeatability, (c) linearity, and (d) dead-band.

5 The following results were obtained from a d.c. tachogenerator test:

Angular velocity (rev/min)	0	500	1000	1500	2000	2500	3000
Output voltage (V)	0	9.1	15.0	23.3	29.9	39.0	47.5

Plot the results and determine (a) the tachogenerator sensitivity from equation 2.1 and (b) the linearity of the device.   [$0.157\,\text{V/(rev/min)}$; 3.6%]

6 State two reasons why the dynamic performance of instruments or measuring systems is important.

7 Determine the time constant τ and the static sensitivity of the systems whose dynamic performances are described by the following differential equations:

a) $30\,\dfrac{d\theta_o}{dt} + 3\theta_o = 1.5 \times 10^{-5}\theta_i$

for a thermocouple in a protective sheath, where θ_o = output voltage in V and θ_i = input temperature in °C.

b) $1.4\,\dfrac{d\theta_o}{dt} + 4.2\theta_o = 9.6\theta_i$

for a restrictor/bellows system, where θ_o = bellows displacement in mm and θ_i = input pressure in bars.   [$10\,\text{s}$; $5 \times 10^{-6}\,\text{V}/°\text{C}$; $0.33\,\text{s}$; $2.29\,\text{mm/bar}$]

8 By examination of the step response of a first-order system shown in fig. 2.7(b), estimate the 5% and 10% settling times expressed in terms of the time constant. Hence calculate the settling times for the systems in question 7.   [30s; 22.5s; 1.0s; 0.75s]

9   A pressure transducer incorporating a bellows and a displacement transducer is a first-order system with a time constant of 0.2s. By referring to the frequency response illustrated in fig. 2.9, determine the maximum frequency of pressure variation that would result in an error of less than 10%.   [0.4Hz]

10   What effects do the values of undamped natural frequency ω_n and damping ratio ζ have on a second-order measuring system? Why is the value of ζ designed to be approximately 0.7 in some recorders?

11   The dynamic performance of a piezo-electric accelerometer can be described by the following differential equation:

$$\frac{d^2\theta_o}{dt^2} + 3.0 \times 10^3 \frac{d\theta_o}{dt} + 2.25 \times 10^{10}\theta_o = 11.0 \times 10^{10}\theta_i$$

where θ_o is the output charge in pC and θ_i is the input acceleration in m/s^2.

By comparison with the standard equations in section 2.3.3, determine the values of ω_n, ζ, and the static sensitivity K. If frequencies up to one fifth of the undamped natural frequency can be measured to within $\pm 5\%$, determine the highest frequency in Hz that can be applied to the accelerometer.   [1.5×10^5 rad/s; 0.01; 4.89 pC/(m/s²); 4.77 kHz]

12   A u.v. galvanometer having an undamped natural frequency ω_n of 15 rad/s and a damping ratio of 0.5 is subjected to a step input. Determine (a) the percentage overshoot, (b) the 5% settling time, (c) the response time, and (d) the rise time.   [17%; 0.15s; 0.17s; 0.1s]

13   A vibration-measuring system gave a 37% overshoot when subjected to a step input. Estimate the amplitude ratio at resonance for the system when subjected to a sine-wave input.   [1.75]

3　Transducers

3.1 Definition
A transducer is a device which converts the quantity being measured into an optical, mechanical, or – more commonly – electrical signal. The energy-conversion process that takes place is referred to as *transduction*.

3.2 Classification
Transducers are classified according to the transduction principle involved and the form of the measurand. Thus a resistance transducer for measuring displacement is classified as a resistance displacement transducer. Other classification examples are pressure bellows, force diaphragm, pressure flapper–nozzle, and so on.

An additional classification could include analogue or digital.

3.3 Transducer elements
Although there are exceptions, most transducers consist of a sensing element and a conversion or control element, as shown in the two-block diagram of fig. 3.1.

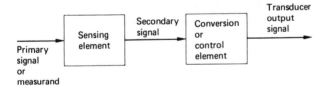

Fig. 3.1 Two-block-diagram representation of a typical transducer

For example, diaphragms, bellows, strain tubes and rings, Bourdon tubes, and cantilevers are sensing elements which respond to changes in pressure or force and convert these physical quantities into a displacement. This displacement may then be used to change an electrical parameter such as voltage, resistance, capacitance, or inductance. Such combinations of mechanical and electrical elements form electro-mechanical transducing devices or transducers. Similar combinations can be made for other energy inputs such as thermal, photo (light), magnetic, and chemical, giving thermoelectric, photoelectric, electromagnetic, and electrochemical transducers respectively.

3.4 Transducer sensitivity

The relationship between the measurand and the transducer output signal is usually obtained by calibration tests and is referred to as the transducer sensitivity K_t,

i.e. $K_t = \dfrac{\text{output-signal increment}}{\text{measurand increment}}$

In practice, the transducer sensitivity is usually known, and, by measuring the output signal, the input quantity is determined from

$$\text{input} = \frac{\text{output-signal amplitude}}{K_t}$$

The following example, in which a spring is used to transduce force into displacement, illustrates the principle involved.

Example 3.1 If the transducing spring shown in fig. 3.2 deflects 0.05 m when subjected to a force of 10 kN, find the input force for an output displacement of 0.075 m.

Fig. 3.2 Loaded spring of example 3.1

Sensitivity $K_t = \dfrac{x}{F} = \dfrac{0.05\,\text{m}}{10\,\text{kN}}$

$\therefore$ input force required for 0.075 m deflection $= 0.075\,\text{m} \times \dfrac{10\,\text{kN}}{0.05\,\text{m}}$

$= 15\,\text{kN}$

3.5 Characteristics of an ideal transducer

The ideal transducer should exhibit the following characteristics.

a) High fidelity – the transducer output waveform shape should be a faithful reproduction of the measurand; i.e. there should be minimum distortion.

b) There should be minimum interference with the quantity being measured; i.e. the presence of the transducer should not alter the measurand in any way.

c) Size. The transducer must be capable of being placed exactly where it is needed.
d) There should be a linear relationship between the measurand and the transduced signal.
e) The transducer should have minimum sensitivity to external effects. Pressure transducers, for example, are often subjected to external effects such as vibration and temperature.
f) The natural frequency of the transducer should be well separated from the frequency and harmonics of the measurand.

3.6 Electrical transducers
Electrical transducers exhibit many of the ideal characteristics. In addition they offer high sensitivity as well as promoting the possibility of remote indication or measurement.

Electrical transducers can be divided into two distinct groups:

a) Variable-control-parameter types, which include

   i) resistance,
  ii) capacitance,
 iii) inductance, and
 iv) mutual-inductance types.

These transducers all rely on an external excitation voltage for their operation.

b) Self-generating types, which include

   i) electromagnetic,
  ii) thermoelectric,
 iii) photoemissive, and
 iv) piezo-electric types.

These all themselves produce an output voltage in response to the measurand input and their effects are reversible. For example, a piezo-electric transducer normally produces an output voltage in response to the deformation of a crystalline material; however, if an alternating voltage is applied across the material, the transducer exhibits the reversible effect by deforming or vibrating at the frequency of the alternating voltage.

3.7 Resistance transducers
Resistance transducers may be divided into two groups, as follows:

a) Those which experience a large resistance change, measured using potential-divider methods. Potentiometers are in this group.
b) Those which experience a small resistance change, measured by bridge-circuit methods. Examples of this group include strain gauges and resistance thermometers. The details of bridge-circuit measuring techniques are discussed in chapter 4.

3.7.1 *Potentiometers*

A linear wire-wound potentiometer consists of a number of turns of resistance wire wound around a non-conducting former, together with a wiping contact which travels over the bare wires. The construction principles are shown in figs 3.3(a) and (b), which indicate that the wiper displacement can be rotary, translational, or a combination of both to give a helical-type motion. The excitation voltage may be either a.c. or d.c., and the output voltage is proportional to the input motion, provided the measuring device has a resistance which is much greater than the potentiometer resistance.

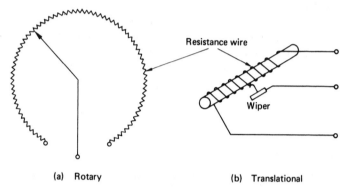

(a) Rotary   (b) Translational

Fig. 3.3 Construction principles of resistance potentiometers

Such potentiometers suffer from the linked problems of resolution and electrical noise. Resolution is defined as the smallest detectable change in input and is dependent on the cross-sectional area of the windings and the area of the sliding contact. The output voltage is thus a series of steps as the contact moves from one wire to the next, as shown in fig. 3.4(a). A conductive plastic potentiometer where the sliding contact is continuous, overcomes this problem.

Electrical noise (i.e. unwanted signals) may be generated by variation in contact resistance, by mechanical wear due to contact friction, and by contact vibration transmitted from the sensing element. In addition, the

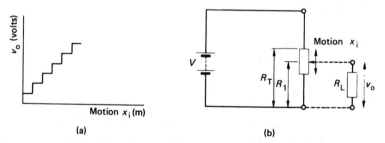

(a)   (b)

Fig. 3.4 Resolution effects and circuit diagram of resistance potentiometer

motion being measured may experience significant mechanical loading by the inertia and friction of the moving parts of the potentiometer. The wear on the contacting surface limits the life of a potentiometer to a finite number of full strokes or rotations, usually referred to in the manufacturer's specification as the 'number of cycles of life expectancy', a typical value being 20×10^6 cycles.

The output voltage v_o of the unloaded potentiometer circuit shown in fig. 3.4(b) is determined as follows.

Let   resistance $R_1 = \dfrac{x_i}{x_T} R_T$

where x_i = input displacement (m)

x_T = maximum possible displacement (m)

R_T = total resistance of the potentiometer (Ω)

then   output voltage $v_o = V \times \dfrac{R_1}{R_1 + (R_T - R_1)}$

$$= V \frac{R_1}{R_T} = V \frac{x_i}{x_T} \times \frac{R_T}{R_T}$$

$$= V \frac{x_i}{x_T} \qquad\qquad 3.1$$

This shows that there is a straight-line relationship between output voltage and input displacement for the unloaded potentiometer.

It would seem that high sensitivity could be achieved simply by increasing the excitation voltage V. However, the maximum value of V is determined by the maximum power dissipation P of the fine wires of the potentiometer winding and is given by

$$V = \sqrt{PR_T} \qquad\qquad 3.2$$

Example 3.2   A potentiometer resistance transducer has a total winding resistance of $10\text{k}\Omega$ and a maximum displacement range of 4cm. If the maximum power dissipation is not to exceed 40mW, determine the output voltage of the device when the input displacement is 1.2cm, assuming the maximum permissible excitation voltage is used.

Using equation 3.2,

excitation voltage $V = \sqrt{PR_T}$

$$= \sqrt{0.04\,\text{W} \times 10000\,\Omega}$$

$$= 20\,\text{V}$$

From equation 3.1,

$$v_o = V \frac{x_i}{x_T}$$

$$= 20\,\mathrm{V} \times \frac{1.2\,\mathrm{cm}}{4\,\mathrm{cm}} = 6\,\mathrm{V}$$

a) Loading a potentiometer

When the potentiometer is loaded by placing across its terminals a measuring device such as a meter, having a resistance R_L, a current flows into the meter. This has a loading effect on the potentiometer and causes the output/input graph to depart from the linear relationship as shown in fig. 3.5.

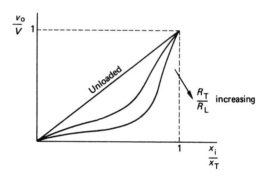

Fig. 3.5 Characteristic of a loaded potentiometer

An analysis of the circuit in the loaded condition gives

$$v_0 = V \left[\frac{x_T}{x_i} + \frac{R_T}{R_L} \left(1 - \frac{x_i}{x_T} \right) \right]^{-1} \tag{3.3}$$

which is far from linear, and the non-linearity increases as the ratio R_T/R_L increases.

Example 3.3 Calculate the error, at 50% full-scale travel of the wiper, of a resistance potentiometer when loaded with a meter having a resistance equal to twice the potentiometer resistance.

Using equations 3.1 and 3.3,

$$\text{unloaded } v_0 = \frac{V}{2} = 0.5\,\mathrm{V}$$

Using equation 3.3,

$$\text{loaded } v_0 = V \left[\frac{1}{2 + \frac{1}{2}(1 - 0.5)} \right]$$

$$= \frac{V}{2.25} = 0.44\,\mathrm{V}$$

32

Hence   error $= \dfrac{0.44\,\text{V} - 0.5\,\text{V}}{0.5\,\text{V}} \times 100\%$

$= -12\%$

(Note the negative sign, which shows that the reading is too low.)

3.7.2 Resistance strain gauges
Resistance strain gauges are transducers which exhibit a change in electrical resistance in response to mechanical strain. They may be of the bonded or unbonded variety.

a) Bonded strain gauges
Using an adhesive, these gauges are bonded, or cemented, directly on to the surface of the body or structure which is being examined.

Examples of bonded gauges are

  i) fine wire gauges cemented to a paper backing,
 ii) photo-etched grids of conducting foil on an epoxy-resin backing,
iii) a single semiconductor filament mounted on an epoxy-resin backing with copper or nickel leads.

As shown in chapter 8, resistance gauges can be made up as single elements to measure strain in one direction only, or a combination of elements such as rosettes will permit simultaneous measurements in more than one direction.

b) Unbonded strain gauges
A typical unbonded-strain-gauge arrangement is shown in fig. 3.6, which shows fine resistance wires stretched around supports in such a way that the deflection of the cantilever spring system changes the tension in the wires and thus alters the resistance of the wire. Such an arrangement may be found in commercially available force, load, or pressure transducers.

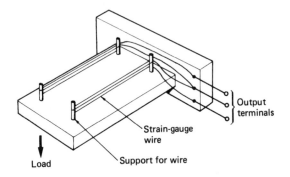

Fig. 3.6 Unbonded strain gauge

3.7.3 Resistance temperature transducers

The materials for these can be divided into two main groups:

a) Metals such as platinum, copper, tungsten, and nickel which exhibit small increases in resistance as the temperature rises; i.e. they have a positive temperature coefficient of resistance.

b) Semiconductors, such as thermistors which use oxides of manganese, cobalt, chromium, or nickel. These exhibit large non-linear resistance changes with temperature variation and normally have a negative temperature coefficient of resistance.

a) Metal resistance temperature transducers

These depend, for many practical purposes and within a narrow temperature range, upon the relationship

$$R_1 = R_0[1 + \alpha(\theta_1 - \theta_0)] \qquad\qquad 3.4$$

where α = temperature coefficient of resistance in $°C^{-1}$

and $\quad R_0$ = resistance in ohms at the reference temperature $\theta_0 = 0°C$

The International Practical Temperature Scale is based on the platinum resistance thermometer, which covers the temperature range $-259.35°C$ to $630.5°C$.

Typical characteristic curves for a platinum resistance thermometer are shown in fig. 3.7.

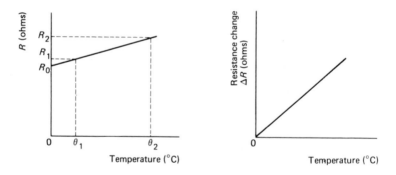

Fig. 3.7 Characteristics of a platinum resistance thermometer

Example 3.4 If the resistance of a platinum resistance thermometer is 100Ω at $0°C$, calculate the resistance at $60°C$ if $\alpha = 0.00392°C^{-1}$.

Using equation 3.4,

$$R_1 = R_0[1 + \alpha(\theta_1 - \theta_0)]$$
$$= 100\Omega \times [1 + 0.00392 \times 60] = 123.5\Omega$$

b) Thermistor (semiconductor) resistance temperature transducers

Thermistors are temperature-sensitive resistors which exhibit large non-linear resistance changes with temperature variation. In general, they have a negative temperature coefficient as illustrated in fig. 3.8.

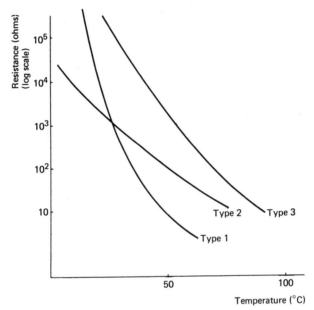

Fig. 3.8 Characteristics of thermistors

For small temperature increments the variation in resistance is reasonably linear; but, if large temperature changes are experienced, special linearising techniques are used in the measuring circuits to produce a linear relationship of resistance against temperature.

Thermistors are normally made in the form of semiconductor discs or beads enclosed in glass envelopes or vitreous enamel. Since they can be made as small as 1 mm, quite rapid response times are possible.

Example 3.5 Use the characteristic curve for the type-1 thermistor shown in fig. 3.8 to determine the temperature measured when the meter in the circuit shown in fig. 3.9 reads half full scale.

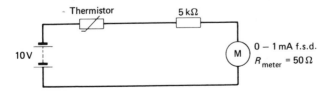

Fig. 3.9 Circuit for example 3.5

35

$$\text{Total resistance } R = \frac{V}{I} = \frac{10\,\text{V}}{0.5 \times 10^{-3}\,\text{A}} = 20\,\text{k}\Omega$$

$$\therefore \quad \text{thermistor resistance} = 20\,\text{k}\Omega - 5\,\text{k}\Omega$$

$$= 15\,\text{k}\Omega \quad (\text{neglecting meter resistance})$$

hence, from the characteristic,

$$\text{temperature} \simeq 20°\text{C}$$

3.7.4 Photoconductive cells

The photoconductive cell, fig. 3.10, uses a light-sensitive semiconductor material. The resistance between the metal electrodes decreases as the intensity of the light striking the semiconductor increases. Common semiconductor materials used for photoconductive cells are cadmium sulphide, lead sulphide, and copper-doped germanium.

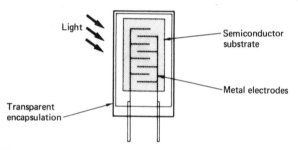

Fig. 3.10 Photoconductive cell

The useful range of frequencies is determined by the material used. Cadmium sulphide is mainly suitable for visible light, whereas lead sulphide has its peak response in the infra-red region and is, therefore, most suitable for flame-failure detection and temperature measurement.

3.7.5 Photoemissive cells (variable conduction or inverse resistance)

When light strikes the cathode of the photoemissive cell shown in fig. 3.11, electrons are given sufficient energy to leave the cathode. The positive anode attracts these electrons, producing a current which flows through resistor R_L and resulting in an output voltage v_o.

$$\text{Photoelectrically generated voltage } v_o = I_p R_L \qquad 3.5$$

where I_p = photoelectric current (A)

and   photoelectric current $I_p = K_t \Phi$

where K_t = sensitivity (A/lm)

and Φ = illumination input (lumen)

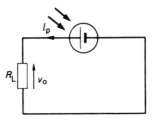

Fig. 3.11 Photoemissive cell

Although the output voltage does give a good indication of the magnitude of illumination, the cells are more often used for counting or control purposes, where the light striking the cathode can be interrupted.

Example 3.6 A photoemissive cell is connected in series with a $5\,k\Omega$ resistor. If the cell has a sensitivity of $30\,\mu A/lm$, calculate the input illumination when the output voltage is $2\,V$.

Using equation 3.5,

$$v_0 = I_p R_L$$
$$\quad = K_t \Phi R_L$$

$$\therefore \quad \text{illumination } \Phi = \frac{v_0}{K_t R_L} = \frac{2\,V}{30 \times 10^{-6}\,A/lm \times 5 \times 10^3\,\Omega} = 13.3\,lm$$

3.8 Capacitive transducers
The capacitance of a parallel-plate capacitor is given by

$$C = \epsilon_0 \epsilon_r \frac{A}{d} \text{ farads} \tag{3.6}$$

where ϵ_0 = the permittivity of free space = $8.854 \times 10^{-12}\,F/m$

      ϵ_r = relative permittivity of the material between the plates

      A = overlapping or effective area between plates (m²)

and d = distance between plates (m)

The capacitance can thus be made to vary by changing either the relative permittivity ϵ_r, the effective area A, or the distance separating the plates d. Some examples of capacitive transducers are shown in fig. 3.12.

The characteristic curves shown in fig. 3.13 indicate that variations of area A and relative permittivity ϵ_r give a linear relationship between C and A or ϵ_r, but variations in spacing d give a linear relationship only over a small range of spacings.

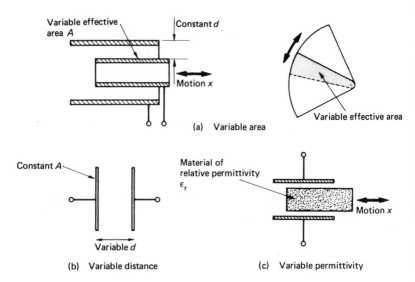

Fig. 3.12 Examples of capacitive transducers

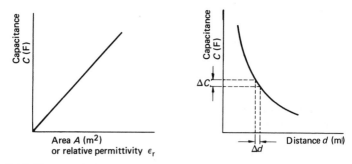

Fig. 3.13 Characteristics of capacitive transducers

By differentiating equation 3.6, we can find the sensitivity in farads/m,

i.e. $$\frac{dC}{dd} = -\frac{\epsilon_0 \epsilon_r A}{d^2}$$ 3.7

Thus the sensitivity is high for small values of d.

Unlike the potentiometer, the variable-distance capacitive transducer has an infinite resolution, making it most suitable for measuring small increments of displacement or quantities which may be changed to produce a displacement.

Example 3.7 A parallel-plate air-spaced capacitor has an effective plate area of $6.4 \times 10^{-4} \text{m}^2$, and the distance between the plates is 1 mm. If the relative permittivity for air is 1.0006, calculate the displacement sensitivity of the device.

Using equation 3.6,

$$C = \epsilon_0 \epsilon_r \frac{A}{d}$$

differentiating,

$$\frac{dC}{dd} = -\frac{\epsilon_0 \epsilon_r A}{d^2}$$

$$= -\frac{8.854 \times 10^{-12} \text{F/m} \times 1.0006 \times 6.4 \times 10^{-4} \text{m}^2}{(1 \times 10^{-3} \text{m})^2}$$

$$= -56.6 \times 10^{-10} \text{F/m}$$

$$= -5.66 \text{nF/m}$$

The minus sign indicates a reduction in the capacitance value for increasing d.

3.9 Inductive transducers
The inductance of a coil wound around a magnetic circuit is given by

$$L = \frac{\mu_0 \mu_r N^2 A}{l} \text{ henrys} \qquad 3.8$$

where μ_0 = permeability of free space = $4 \times 10^{-7} \text{H/m}$

        μ_r = relative permeability

        N = number of turns on coil

        l = length of magnetic circuit (m)

and A = cross-sectional area of magnetic circuit (m^2)

This can be rewritten as

$$L = \frac{N^2}{S} \qquad 3.9$$

where S is the magnetic reluctance of the inductive circuit.

The inductance can thus be made to vary by changing the reluctance of the inductive circuit. Some examples of variable-reluctance transducers are shown in figs 3.14 (a) to (c).

A typical characteristic curve for an inductive transducer is shown in fig. 3.15.

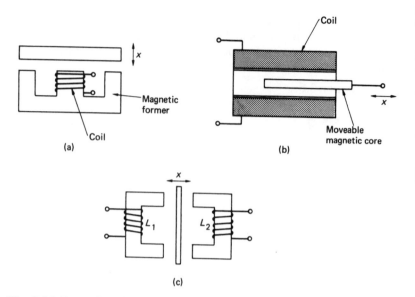

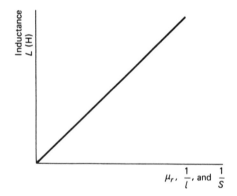

Fig. 3.14 Examples of variable-reluctance inductive transducers

Fig. 3.15 Characteristic of inductive transducer

Example 3.8   Determine the sensitivity of a single-coil inductive transducer for (a) variations in relative permeability μ_r, (b) variations in length of magnetic circuit.

a) Differentiating equation 3.8 with respect to μ_r,

$$\frac{\mathrm{d}L}{\mathrm{d}\mu_r} = \frac{\mu_0 N^2 A}{l}$$

b) Differentiating equation 3.8 with respect to l,

$$\frac{dL}{dl} = - \frac{\mu_0 \mu_r N^2 A}{l^2}$$

(note – high for small values of l)

Measuring techniques used with capacitive and inductive transducers

a) A.C.-excited bridges using differential capacitors or inductors.
b) A.C. potentiometer circuits for dynamic measurements.
c) D.C. circuits to give a voltage proportional to velocity for a capacitor.
d) Frequency-modulation methods, where the change of C or L varies the frequency of an oscillating circuit.

Important features of capacitive and inductive transducers are as follows:

  i) Resolution   infinite
 ii) Accuracy $\pm 0.1\%$ of full scale is quoted
iii) Displacement ranges 25×10^{-6} m to 10×10^{-3} m
 iv) Rise time   less than $50\,\mu s$ possible

Typical measurands are displacement, pressure, vibration, sound, and liquid level.

3.10 Linear variable-differential transformer (l.v.d.t.)
A typical differential transformer, as illustrated in fig. 3.16, has a primary coil, two secondary coils, and a movable magnetic core.

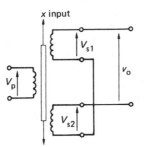

Fig. 3.16 Details of an l.v.d.t.

A high-frequency excitation voltage V_p is applied to the primary winding and, due to transformer action, voltages V_{s1} and V_{s2} are induced in the secondary coils. The amplitudes of these secondary voltages are dependent on the degree of electromagnetic coupling between the primary and secondary coils and hence on the core displacement x.

Since the secondary coils are connected in series opposition, the displacement x of the core which produces an increase in V_{s1} will produce a

41

corresponding decrease in V_{s2}. Ideally the voltages V_{s1} and V_{s2} should be 180° out of phase with each other, so that at the central position there is zero output voltage. However, the voltages generally are not exactly 180° out of phase and there is a small null output voltage as illustrated in fig. 3.17(b).

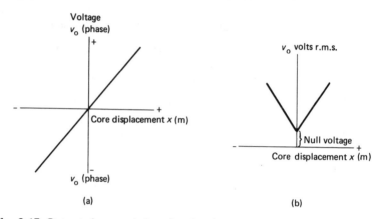

Fig. 3.17 Output characteristics of an l.v.d.t.

Some important characteristics and features of the l.v.d.t. are as follows:

a) infinite resolution;
b) linearity better than 0.5%;
c) excitation frequency 50 Hz to 20 kHz;
d) null voltage less than 1% of full-scale output voltage;
e) maximum displacement frequency 10% of the excitation frequency;
f) displacement ranges available from 2×10^{-4} m to 0.5 m;
g) no wear of moving parts;
h) amplitude-modulated output, i.e. the output voltage is a constant-frequency waveform with an amplitude depending on the displacement input.

Typical measurands are any quantities which can be transduced into a displacement, e.g. pressure, acceleration, vibration, force, and liquid level.

3.11 Piezo-electric transducers
When a force is applied across the faces of certain crystal materials, electrical charges of opposite polarity appear on the faces due to the piezo-electric effect ('piezo' comes from the Greek for 'to press'). Piezo-electric transducers are made from natural crystals such as quartz and Rochelle salt, synthetic crystals such as lithium sulphate, or polarised ceramics such as barium titanate. Since these materials generate an

42

output charge proportional to applied force, they are most suitable for measuring force-derived variables such as pressure, load, and acceleration as well as force itself.

Piezo-electric materials are good electrical insulators; therefore, with their connecting plates, they can be considered as parallel-plate capacitors as shown in fig. 3.18(a). When a force is applied, the capacitor simply 'charges up' due to the piezo-electric effect, as illustrated by the equivalent electric circuit shown in fig. 3.18(b). Unfortunately, any measuring instrument electrically connected across the capacitor C will tend to discharge it; hence the transducer's steady-state response is poor. This can be overcome by using measuring amplifiers with very high input impedances (10^{12} to 10^{14} ohms being typical) known as charge amplifiers (see chapter 4), but these make the measuring system increasingly expensive.

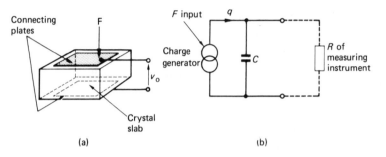

Fig. 3.18 Piezo-electric transducer

Example 3.9 A piezo-electric pressure transducer has a sensitivity of 80 pC/bar. If it has a capacitance of 1 nF, determine its output voltage when the input pressure is 1.4 bar.

Charge q = sensitivity × pressure

$$= 80 \frac{\text{pC}}{\text{bar}} \times 1.4 \text{bar}$$

$$= 80 \times 1.4 \text{pC} = 112 \text{pC}$$

Output voltage $V = \dfrac{q}{C} = \dfrac{112 \times 10^{-12}\text{C}}{1 \times 10^{-9}\text{F}}$

$$= 112 \text{mV}$$

3.12 Electromagnetic transducers
These employ the well known generator principle of a coil moving in a magnetic field.

The output voltage of the electromagnetic transducer is given as follows.

a) For a coil with changing flux linkages,

$$\text{output voltage } v_\text{o} = -N\frac{d\Phi}{dt}$$

where N = number of turns on coil

and $\dfrac{d\Phi}{dt}$ = rate at which flux changes (Wb/s)

b) For the single conductor moving in a magnetic field,

output voltage $v_\text{o} = Blv$

where B = flux density (T)

l = length of conductor (m)

and v = velocity of conductor perpendicular to flux direction (m/s)

Both relationships are used in commercially available velocity transducers, the construction principles of which are illustrated in figs 3.19 (a) to (c).

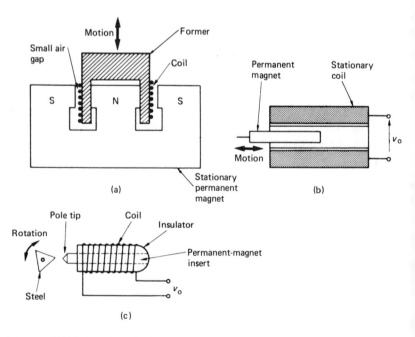

Fig. 3.19 Electromagnetic transducers

Some important features of the electromagnetic transducer are as follows:

i) output voltage is proportional to the velocity of input motion;
ii) usually they have a large mass, hence they tend to have low natural frequencies;
iii) high power outputs are available;
iv) limited low-frequency response – ranges from 10 Hz to 1 kHz are quoted in manufacturers' literature.

3.13 Thermoelectric transducers

When two dissimilar metals or alloys are joined together at each end to form a thermocouple as shown in fig. 3.20 and the ends are at different temperatures, an e.m.f. will be developed causing a current to flow around the circuit. The magnitude of the e.m.f. depends on the temperature difference between the two junctions and on the materials used. This thermoelectric effect is known as the Seebeck effect and is widely used in temperature-measurement and control systems.

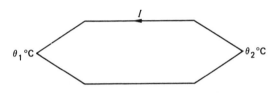

Fig. 3.20 Thermocouple circuit

The main problems with thermocouples are corrosion, oxidation, or general contamination by the atmosphere of their location. These problems can be overcome by the selection of a protective sheath which does not react with the atmosphere or fluid.

Although they do give a direct output voltage, this is generally small – in the order of millivolts – and often requires amplification.

Advantages of thermocouples include

a) temperature at localised points can be determined, because of the small size of the thermocouple;
b) they are robust, with a wide operating range from $-250°C$ to $2600°C$.

3.14 Photoelectric cells (self-generating)

The photoelectric or photovoltaic cell makes use of the photovoltaic effect, which is the production of an e.m.f. by radiant energy – usually light – incident on the junction of two dissimilar materials. The construction of a typical cell is illustrated in fig. 3.21(a), which shows a sandwich layer of metal, semiconductor material, and a transparent layer. Light travelling through the transparent layer generates a voltage

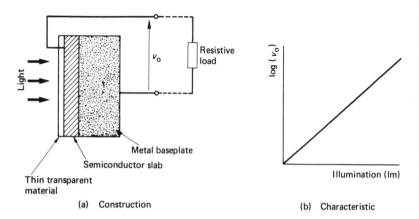

Fig. 3.21 Photoelectric cell

which is a logarithmic function of light intensity. The device is highly sensitive; has a good frequency response; and, because of its logarithmic relationship of voltage against light, is very suitable for sensing over a wide range of light intensities. The characteristic of the device is shown in fig. 3.21(b).

3.15 Mechanical transducers and sensing elements
Many transducing systems consist of two different types of transducer operating in series, or cascade. In the electrical-transducer section it was assumed that the input to the transducer was provided by the sensing element. The sensing element itself is often a mechanical transducer which converts the measurand into a displacement or force which is then used to change some electrical parameter.

Some of the more common mechanical transducers are shown in the following examples.

3.15.1 Force-to-displacement transducers

a) Spring
The spring shown in example 3.1 (fig. 3.2) is the simplest form of mechanical transducer. For equilibrium, we have

$$F = \lambda x$$

where λ = spring stiffness (N/m)

$$\therefore \quad x = \frac{F}{\lambda} \quad \text{or} \quad \lambda = \frac{F}{x}$$

But sensitivity $K_t = \dfrac{x}{F}$ (see example 3.1)

$$\therefore \quad \text{sensitivity} = \frac{1}{\lambda}$$

i.e. the stiffer the spring, the smaller the sensitivity.

b) Cantilever
When the cantilever shown in fig. 3.22 is loaded, it experiences a deflection y. The relationship between the force F and the deflection is given by

deflection y = constant × force

$$\therefore \qquad y = kF$$

where the constant k depends on the material and dimensions of the cantilever.

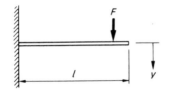

Fig. 3.22 Cantilever

3.15.2 Pressure-to-displacement transducers

a) Diaphragms
Pressure can be measured using a steel diaphragm as shown in fig. 3.23. The displacement x of the diaphragm is proportional to the pressure

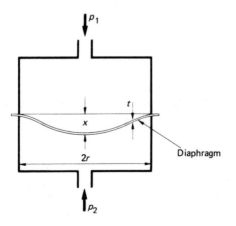

Fig. 3.23 Diaphragm

47

difference $(p_1 - p_2)$ if the displacement is less than one third of the diaphragm thickness t. The relationship between pressure differential $(p_1 - p_2)$ and diaphragm displacement is thus given by

deflection = constant × pressure differential

$$\therefore \qquad x = k(p_1 - p_2)$$

where k depends on the material and dimensions of the diaphragm.

The diaphragm, usually a thin flat plate of spring steel, may be used with an electrical transducer to produce a small transducer having high sensitivity.

b) Bourdon tubes
This type of transducer, illustrated in fig. 3.24, is used in many commercially available pressure gauges. The main feature of Bourdon tubes is their large deflection.

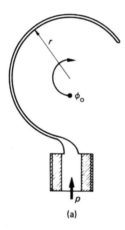

(b)   Tube cross-section

(a)

Fig. 3.24 Bourdon tube

Provided the major axis a of the cross-section is considerably larger than the minor axis b, the following relationship holds between the input pressure p and the tube-tip deflection ϕ_o:

$$\phi_o = \text{constant} \times p$$

c) Bellows
This is basically a pneumatic spring, as illustrated in fig. 3.25, and is in general use in pneumatic instruments.

Equating the forces acting on the bellows, for equilibrium

$$pA = \lambda x$$

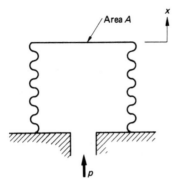

Fig. 3.25 Bellows

$$\therefore \quad x = \frac{A}{\lambda}p$$

where A = cross-sectional area of bellows (m²)
        p = input pressure (N/m²)
and λ = bellows stiffness (N/m)

3.15.3 Displacement-to-pressure transducers

A typical arrangement of a flapper–nozzle system is shown in fig. 3.26(a).

For a constant supply pressure p_s, movement of the flapper allows a variable bleed-off of air which varies the control pressure p_c. The transducer characteristic, illustrated in fig. 3.26(b), is non-linear but has a narrow linear region which is usually extended by employing feedback when the transducers are used in pneumatic instruments. The flow rates used are very small, and a pneumatic amplifier must be employed to boost the controlled pressure.

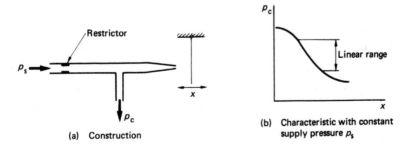

(a) Construction

(b) Characteristic with constant supply pressure p_s

Fig. 3.26 Flapper–nozzle system

Exercises on chapter 3

1 The output voltage of a potentiometer-type resistance transducer is to be measured by a recorder having an input resistance of 20 kΩ. If the error of measurement is not to exceed −2% at 50% f.s.d., determine the resistance value of the potentiometer. [1.633 kΩ]

2 The following is a typical specification for a potentiometer-type resistance transducer. Examine the specification and explain the meaning and significance of each item.

Type   wire-wound resistance displacement potentiometer

Terminal resistance	10 kΩ	Range	0–25 mm
Resolution	0.4%	Power rating	0.25 W
Maximum wiper current	15 mA	Thermal drift	0.05% per °C
Life expectancy	10^8 cycles		

3 In the circuit shown in fig. 3.27, the relay RL 1 operates at 6 V and has a resistance of 1 kΩ. Using the type-2 thermistor characteristic shown in fig. 3.8, determine the thermistor temperature when the relay operates. [25°C]

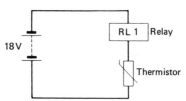

Fig. 3.27 Circuit diagram for exercise 3

4 A linear variable-differential transformer is excited with a 100 Hz 6 V peak-to-peak waveform. The input core motion is sinusoidal at 10 Hz and has a displacement amplitude of ±3 mm. If the l.v.d.t. sensitivity is 2 V/mm, draw the waveforms of the excitation voltage, input displacement, and output voltage.

5 The specification for the l.v.d.t. in question 4 is as follows:

Linearity	0.4%	Resolution	infinite
Residual voltage	0.5%	Drift	better than 0.1% per °C
Output impedance	2.5 kΩ	Response time	1 ms

Explain the meaning and significance of the specification.

6 (a) Describe the principle of operation and construction details of a piezo-electric (quartz) transducer.

(b) A quartz pressure transducer has a sensitivity of 80 pC/bar. If, when the input pressure is 3 bars, an output voltage of 1 V is produced, determine the capacitance of the device. [240 pF]

7 Describe the principle of operation of photoconductive and photovoltaic transducers and, with the aid of simple sketches, describe a simple engineering application for each one.

4 Signal conditioning

4.1 Introduction
The transduced signal is rarely in a form ready for display or recording – it may need to be increased in magnitude or modified in some way before display. The process of preparing the signal before display or recording is referred to as signal conditioning.

The signal conditioner may have one or all of the following functions:

a) *Amplification* The small signal from the transducer is increased in magnitude by a device referred to as an amplifier, e.g. levers, gears, and electronic, pneumatic, and hydraulic amplifiers. The amount by which the signal is increased in magnitude is referred to as either gain, amplification, or magnification.

b) *Signal modification* The form of the signal or amplified signal is changed, e.g. by rack-and-pinion gears, electronic modulators, bridge circuits, potentiometric circuits, and analogue-to-digital convertors. The distinction between transducers and signal modifiers is not always clear, and for our purposes it will be assumed that the transducer both senses and modifies the measurand.

c) *Impedance matching* The signal conditioner acts as a buffer stage between the transducing and recording elements, the input and output impedances of the matching device being arranged to prevent loading of the transducer and maintain a high signal level at the recorder.

4.2 Amplifiers
An amplifier is a device which increases the magnitude of, or amplifies, its input signal.

Consider the block-diagram representation of an amplifier shown in fig. 4.1. The input signal θ_i is amplified by an amount G, resulting in an output θ_o which is given by

$$\theta_o = G\theta_i$$

$$\therefore \quad \frac{\theta_o}{\theta_i} = G, \quad \text{the gain or amplification} \tag{4.1}$$

Since θ_o and θ_i are in the same units, the ratio G is therefore dimensionless.

Fig. 4.1 An amplifier block diagram

Example 4.1 A displacement magnifier has an amplification of 20000. If the output displacement is 4 cm, determine the corresponding input displacement.

Using equation 4.1,

i.e. $\dfrac{\theta_o}{\theta_i} = G$

$$\theta_i = \frac{\theta_o}{G} = \frac{4 \times 10^{-2}\,\text{m}}{20000} = 2 \times 10^{-6}\,\text{m}$$

4.2.1 Amplifiers in cascade

If an amplifier having a gain G_1 is so arranged that its output signal becomes the input signal of another amplifier having a gain G_2, as shown in fig. 4.2, the two amplifiers are said to be *in cascade* and the overall gain or amplification of the combined devices is

$$\frac{\theta_o}{\theta_i} = G_1 G_2 \qquad\qquad 4.2$$

i.e. the product of their individual gains, assuming no loading occurs.

Fig. 4.2 Amplifiers in cascade

Example 4.2 Two amplifiers A and B are cascaded so that their combined gain is the product of their individual gains. Given that gain of amplifier A = 100 and gain of amplifier B = 300, determine the output produced by an input of 4 units.

Using equation 4.2,

i.e. $\dfrac{\theta_o}{\theta_i} = G_1 G_2$

$\theta_o = \theta_i G_1 G_2$

$\quad = 4\ \text{units} \times 100 \times 300$

$\quad = 120000\ \text{units}$

4.3 Mechanical amplifiers

4.3.1 The simple lever

Levers and gears are used as displacement amplifiers in instruments such as dial-test indicators, extensometers, and pressure gauges.

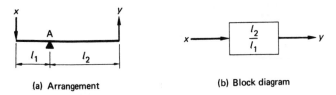

(a) Arrangement   (b) Block diagram

Fig. 4.3 The simple lever

Consider fig. 4.3(a), which shows a simple lever pivoted at A. If the input is a displacement x, the lever will cause the output end to displace by an amount y. By similar triangles, the displacement gain or amplification is given by

$$\frac{y}{x} = \frac{l_2}{l_1}$$

i.e. $\quad y = \frac{l_2}{l_1} x$ 　　　　　　　　　　　　　4.3

which can be represented by the block diagram shown in fig. 4.3(b).

Note that the input and output displacements are of opposite phase, i.e. if x is 'down' then y is 'up'.

4.3.2 The compound lever

The compound lever shown in fig. 4.4(a) overcomes the problem of phase reversal of the displacement signal as well as increasing the amplification.

The two levers are linked together so that the output of one lever provides the input to the other.

Now $\quad \dfrac{z}{y} = \dfrac{l_4}{l_3}$

but $\quad y = x\dfrac{l_2}{l_1}$ (from equation 4.3)

hence the overall displacement gain or amplification is

$$\frac{z}{x} = \frac{l_4}{l_3} \times \frac{l_2}{l_1}$$ 　　　　　　　4.4

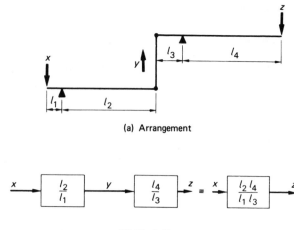

(a) Arrangement

(b) Block diagram

Fig. 4.4 The compound lever

and the output is

$$z = \frac{l_4}{l_3} \times \frac{l_2}{l_1} \times x$$

which can be represented by the block diagram shown in fig. 4.4(b).

Example 4.3 A compound lever consists of two levers A and B. Lever A has a length magnifying ratio of 2.8:1, and lever B has a length magnifying ratio of 6:1. Determine the displacement gain of the device.

Using equation 4.4,

$$\text{Gain} = \frac{l_4}{l_3} \times \frac{l_2}{l_1}$$

$$= \frac{2.8}{1} \times \frac{6}{1} = 16.8$$

i.e. the compound lever has a displacement gain of 16.8.

4.3.3 The Huggenberger extensometer

The Huggenberger extensometer uses the amplifying-lever principle to measure the extension of a specimen used in a tensile test. It is one of the most popular and accurate mechanical gauges in use today, with a typical construction as shown in fig. 4.5 producing displacement magnifications of up to 2000.

The magnification factor can be found as follows. Let the moveable knife-edge move a distance Δx; then

54

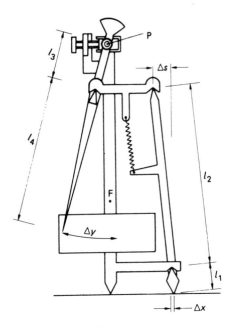

Fig. 4.5 Huggenberger extensometer

$$\Delta s = \frac{l_2}{l_1} \Delta x$$

and the pointer movement is

$$\Delta y = \frac{l_4 + l_3}{l_3} \Delta s = \frac{l_4 + l_3}{l_3} \times \frac{l_2}{l_1} \Delta x$$

$$\therefore \quad \text{magnification} = \frac{\Delta y}{\Delta x} = \frac{l_2(l_4 + l_3)}{l_1 l_3}$$

The device can be reset during a series of measurements by means of the screw which adjusts the pivot point P without disturbing the rest of the linkage system. The gauge is mounted by a clamp, springs, or screw pressure applied to the frame F to set the knife-edges on to the specimen. The application of the device is severely limited by this clamping arrangement and by the height of the gauge, which makes it very unstable in position.

4.3.4 Simple gears
Simple gears may be used to provide magnification of either angular displacement or rotational speed.

Consider the two meshed gears shown in fig. 4.6(a). If θ_o is the angular displacement of the output gear and θ_i the angular displacement of the

55

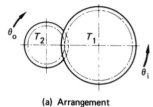

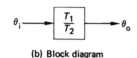

(b) Block diagram

(a) Arrangement

Fig. 4.6 The simple gear

input gear, and T_2 and T_1 are their respective number of teeth, then the radial-displacement amplification is given by

$$\frac{\theta_o}{\theta_i} = \frac{T_1}{T_2}$$

hence $\quad \theta_o = \frac{T_1}{T_2} \times \theta_i$

or, if rotational speed N is being considered,

$$N_2 = \frac{T_1}{T_2} \times N_1 \tag{4.5}$$

This system can be represented by the block diagram shown in fig. 4.6(b). Note the change of direction between input and output.

4.3.5 Compound gears

A typical compound-gear arrangement using four gears is shown in fig. 4.7(a). If the gears have T_1, T_2, T_3, and T_4 teeth respectively, the relationship between the input θ_i and the output θ_o is given by

$$\frac{\theta_o}{\theta_i} = \frac{T_1}{-T_2} \times \frac{T_3}{T_4} \tag{4.6}$$

with no change in direction, which can be represented by the block diagram shown in fig. 4.7(b).

Example 4.4 If, in the compound-gear arrangement shown in fig. 4.7(a), $T_1 = 60$ teeth, $T_2 = 12$ teeth, $T_3 = 40$ teeth, and $T_4 = 25$ teeth, determine the gain of the compound system.

Using equation 4.6,

$$\text{Gain} = \frac{T_1}{T_2} \times \frac{T_3}{T_4}$$

$$= \frac{60}{12} \times \frac{40}{25} = 8$$

i.e. the system gives a displacement amplification or gain of 8.

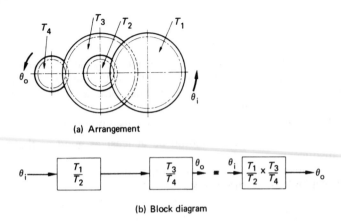

(a) Arrangement

$$\theta_i \longrightarrow \boxed{\dfrac{T_1}{T_2}} \longrightarrow \boxed{\dfrac{T_3}{T_4}} \xrightarrow{\theta_o} \quad \equiv \quad \xrightarrow{\theta_i} \boxed{\dfrac{T_1}{T_2} \times \dfrac{T_3}{T_4}} \longrightarrow \theta_o$$

(b) Block diagram

Fig. 4.7 The compound gear

4.3.6 Optical lever

The laws of reflection are used in many electrical instruments such as the u.v. galvanometer recorder which is examined in detail in chapter 5. The scale of the mechanical-pointer galvanometer may be enlarged if the pointer is made correspondingly long. However, an increase in pointer length increases its inertia, which in turn may mechanically load the sensitive movement. A light source is therefore employed and, provided a long optical arm is used (i.e. a long ray distance from source to scale), large scales may be used.

Consider the rotatable mirror and light source shown in fig. 4.8(a). There is an angle magnification of 2, since the total angle of deflection of the reflected beam is twice the angle of deflection of the rotatable mirror.

The device can be represented by the single block diagram shown in fig. 4.8(b), where

$$\theta_o = 2\theta_i \quad \text{or} \quad \frac{\theta_o}{\theta_i} = 2$$

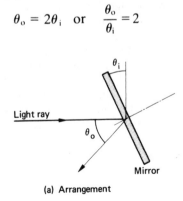

(a) Arrangement

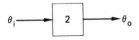

(b) Block diagram

Fig. 4.8 The optical lever

57

4.4 Electronic amplifiers

4.4.1 Introduction

The small-amplitude electrical signals from many electrical transducers are often too small to be applied directly to the display or recording device. In such cases, electronic amplifiers are used to increase the magnitude of the signals.

Three desirable characteristics of electronic amplifiers are

i) the frequency response should be at least as good as that of the transducer;
ii) to minimise the loading effect on the transducer, the amplifier should have a high input impedance;
iii) the amplifier should have a low output impedance, so that the recording device does not load the amplifier.

4.4.2 Types of amplifier

Two types of amplifier used extensively in instrumentation systems are

i) a.c.-coupled amplifiers;
ii) d.c. or directly coupled amplifiers.

The differences between a.c. and d.c. amplifiers may be explained by an examination of their frequency responses, i.e. the graphs of gain or amplification versus frequency.

Figure 2.15 shows the frequency response of an a.c. amplifier which has a constant gain over a range of frequencies between f_1 and f_2. Since there is no response at zero and low frequencies, the amplifier is incapable of handling steady-state (zero frequency) and very-low-frequency signals.

The response of the d.c. amplifier, shown in fig. 2.16, indicates that the d.c. amplifier can respond to signals down to zero frequency and is therefore, unlike its a.c. counterpart, capable of handling steady-state and very low frequencies in addition to the higher frequencies.

Although d.c. amplifiers can respond to signals down to zero frequency, their use may be limited by 'drift' encountered at low signal levels. 'Drift' is the name given to the slow variations in the d.c. voltage available at the amplifier's output terminals and may be due to variations in power-supply voltage, circuit components, transistor characteristics, or other devices used within the amplifier. Fortunately, due to rapid developments within the electronics industry, this problem is not as serious as it used to be, and integrated-circuit technology has now made available inexpensive d.c. amplifiers with very low drift characteristics.

a) Decibel notation applied to an amplifier

In appendix C it is shown that

$$dB = 10\log_{10}\frac{P_2}{P_1} \qquad\qquad 4.7$$

where P_2/P_1 is the ratio of two powers.

In an electronic amplifier, P_2 is the power output in watts and P_1 is the power input in watts. If we assume the input and output resistances are of the same value, then

$$dB = 20\log_{10}\frac{V_2}{V_1} \qquad\qquad 4.8$$

Example 4.5 The power output from an amplifier is increased 4 times. Express this power increase in dB.

Using equation 4.7,

$$power\ increase = 10\log_{10}\frac{P_2}{P_1}$$

$$= 10\log_{10}\frac{4}{1}$$

$$= 6\,dB$$

Example 4.6 A frequency-response test on an a.c. amplifier gave the following results. Calculate the corresponding dB values, reference to 800 Hz, i.e. using the output voltage of 2.12 V at a frequency of 800 Hz as the reference voltage V_1 in equation 4.8.

Frequency (Hz)	50	100	200	400	800	1600	3200	6400
Output voltage (V (r.m.s.))	1	1.5	2	2.1	2.12	2.2	2.02	1.47

Using equation 4.8 and a reference of 2.12 V, the corresponding values of dB are

$$-6.53, \quad -3, \quad -0.506, \quad -0.08, \quad 0, \quad 0.32, \quad -0.42, \quad -3.18$$

Note:
i) $-3\,dB$ corresponds to a 29.3% voltage-amplitude reduction;
ii) a graph of dB against frequency in Hz would normally be plotted on linear–log paper.

b) Bandwidth
The bandwidth of an amplifier is often expressed as the frequency range, in Hz, between the '3dB down points', i.e. frequencies at which the voltage output amplitude falls by 29.3% to 70.7% of the maximum value. Thus in fig. 2.15 the bandwidth is given by:

$$bandwidth = (f_2 - f_1)\,Hz \qquad\qquad 4.9$$

Example 4.7 Determine the approximate bandwidth of the amplifier whose frequency response was shown in example 4.6.

−3 dB occurs at 100 Hz and about 6400 Hz. Using the relationship shown in equation 4.9,

$$\text{bandwidth} = (f_2 - f_1)\,\text{Hz}$$
$$= (6400\text{–}100)\,\text{Hz}$$
$$= 6.3\,\text{kHz}$$

4.4.3 The differential amplifier

The differential amplifier shown in fig. 4.9 amplifies the difference between two input signals and is thus described as having a 'double-ended input'. If the input voltages are v_1 and v_2 and the differential voltage gain is G, then the output voltage is given by

$$\text{output voltage } v_o = G(v_1 - v_2) \qquad\qquad 4.10$$

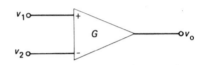

Fig. 4.9 The differential amplifier

Example 4.8 A differential amplifier having a differential voltage gain of 1000 is used to amplify the output voltage from a pressure transducer. Determine the output voltage of the amplifier when (a) input $v_1 = +5\,\text{V}$ and input $v_2 = +5\,\text{V}$; (b) input $v_1 = +5\,\text{V}$ and input $v_2 = +5.001\,\text{V}$.

a) Using equation 4.10,

$$v_o = G(v_1 - v_2)$$
$$= 1000(5 - 5)\,\text{V} = 0\,\text{V}$$

b) Again using equation 4.10,

$$v_o = 1000(5 - 5.001)\,\text{V}$$
$$= -1\,\text{V}$$

Note that no output voltage is produced if the input voltages are of the same polarity and magnitude.

4.4.4 Operational amplifiers

An operational amplifier is a d.c. differential amplifier and associated external components that together 'operate' upon a direct voltage or current in some mathematical way. It is used in a number of applications in instrumentation and control engineering, and its usefulness depends upon the following properties which it possesses:

a) high gain, 200000 to 10^6;
b) phase reversal, i.e. the output voltage is of opposite sign to the input;
c) high input impedance.

Two very important implications of these properties are:

i) With high voltage gain, for any *sensible* output voltage the input voltage will be so small that it may be assumed to be virtually zero. Since the input point is virtually zero volts, it is referred to as a *virtual earth* ('virtual' because it is not connected directly to earth).
ii) Since the input impedance is high and the input voltage is a virtual earth, the amplifier takes negligible current which is therefore assumed to be zero for a simplified analysis.

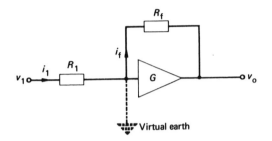

Fig. 4.10 The operational amplifier

Consider fig. 4.10, which shows an operational amplifier in circuit with two resistors R_1 and R_f. Since the input current to the amplifier is negligible,

$$i_1 = i_f$$

i.e. $$\frac{v_1 - 0}{R_1} = \frac{0 - v_o}{R_f}$$

$$\therefore \quad \frac{v_o}{v_1} = -\frac{R_f}{R_1}$$

or $$v_o = -v_1 \frac{R_f}{R_1} \quad \text{(the minus sign indicates reverse polarity)}$$

i.e. the output/input relationship is independent of the amplifier gain and depends only on the resistors R_1 and R_f.

4.4.5 The summer amplifier
Figure 4.11 shows an operational amplifier connected to resistors in such a way that the output voltage v_o may be the sum of the input voltages v_1, v_2, and v_3. Such a summer amplifier may be used as a comparison element in control-engineering applications, as discussed in chapter 14.

61

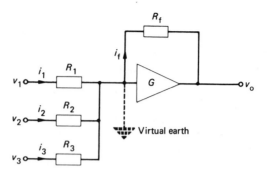

Fig. 4.11 The summer amplifier

Since the input current to the amplifier is zero.

$$i_f = i_1 + i_2 + i_3$$

$$\frac{-v_o}{R_f} = \frac{v_1}{R_1} + \frac{v_2}{R_2} + \frac{v_3}{R_3}$$

$$\therefore \quad v_o = - \left(R_f \frac{v_1}{R_1} + R_f \frac{v_2}{R_2} + R_f \frac{v_3}{R_3} \right)$$

$$= -R_f \left(\frac{v_1}{R_1} + \frac{v_2}{R_2} + \frac{v_3}{R_3} \right) \qquad 4.11$$

and, if $R_1 = R_2 = R_3 = R_f$

$$v_o = -(v_1 + v_2 + v_3)$$

Example 4.9 If in fig. 4.11, $R_1 = 1\,k\Omega$, $R_2 = 2\,k\Omega$, $R_3 = 1.5\,k\Omega$, $R_f = 10\,k\Omega$, $v_1 = 1.5\,V$, $v_2 = 2\,V$, and $v_3 = 3\,V$, determine the output voltage v_o.

Using equation 4.11,

$$v_o = -R_f \left(\frac{v_1}{R_1} + \frac{v_2}{R_2} + \frac{v_3}{R_3} \right)$$

$$= -10\,k\Omega \left(\frac{1.5\,V}{1\,k\Omega} + \frac{2\,V}{2\,k\Omega} + \frac{3\,V}{1.5\,k\Omega} \right)$$

$$= -10\,k\Omega \times 4.5\,V/k\Omega$$

$$= -45\,V$$

4.4.6 *Negative feedback in electronic amplifiers*

Voltage negative feedback is used beneficially in electronic amplifiers to improve their overall performance. The effects of voltage negative feedback include

a) a reduction in gain,
b) improved frequency response,
c) increased bandwidth,
d) increased input impedance,
e) reduction in output impedance,
f) ensuring that the amplifier is less sensitive to component, device-characteristic, and power-supply changes.

a) *Effect of negative feedback on gain*

Consider the amplifier shown in fig. 4.12, having a gain A with a fraction β of the output v_0 fed back in anti-phase (i.e. 180° out of phase) with the input signal. The input to the amplifier is now given by

amplifier input $= v_i - \beta v_0$

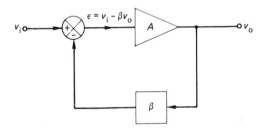

Fig. 4.12 An electronic amplifier with negative feedback

But   amplifier output $=$ amplifier input $\times A$

$\therefore \qquad\qquad v_0 = A(v_i - \beta v_0)$

and $\qquad v_0(1 + A\beta) = A v_i$

Hence, with negative feedback,

$$\text{gain} = \frac{v_0}{v_i} = A' = \frac{A}{1 + A\beta} \qquad\qquad 4.12$$

i.e. the gain A is reduced by $\dfrac{1}{1 + A\beta}$

b) *Effect of negative feedback on frequency response*

If $\quad A\beta \gg 1 \quad$ then $\quad A' = \dfrac{1}{\beta}$

Since the gain A depends solely on β, which does not change with frequency, it will remain constant over a range of frequencies providing the condition $A\beta \gg 1$ exists.

Consider fig. 4.13, which shows a typical a.c. amplifier frequency response, with and without negative feedback. The graph shows a reduction in gain when negative feedback is applied but, since the reduced gain remains constant over a wide frequency range, there is a resultant increase in bandwidth.

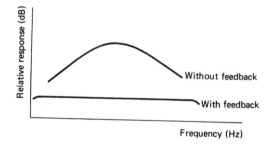

Fig. 4.13 Effects of negative feedback on the frequency response of an amplifier

c) Other effects of negative feedback

When negative feedback is applied, the actual voltage ϵ into the amplifier is reduced. The corresponding reduction in input current, for a given input signal v_i, can be explained by an apparent increase in the input impedance of the amplifier. This effect, along with effects (e) and (f) above can be proved analytically, but it is beyond the scope of this book to do so.

Example 4.10 An amplifier has a voltage gain without feedback of 10^5. If one tenth of the output signal is fed back in anti-phase to the input signal, determine the voltage gain with negative feedback.

Using equation 4.12,

$$A' = \frac{A}{1 + A\beta} = \frac{10^5}{1 + 10^4} = 9.999$$

Alternatively,

$$A' = \frac{1}{\beta} = \frac{1}{0.1} = 10$$

4.5 Signal modifiers or converters

The transduced or amplified signal may require further conditioning before recording or display. This may be accomplished using a signal modifier or converter such as

a) a rack-and-pinion gear,
b) a charge amplifier,
c) a modulation system,
d) a bridge circuit, or
e) analogue to digital and digital to analogue converters.

4.5.1 *Rack-and-pinion gear*
The rack-and-pinion gear shown in fig. 4.14 converts linear motion into rotational motion and vice-versa. This device, with a set of gears, is used in the dial-test indicator discussed in chapter 6.

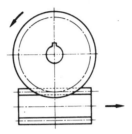

Fig. 4.14 Rack-and-pinion gear

4.5.2 *The charge converter*
The charge generated by a piezo-electric crystal transducer may be converted into a voltage by a charge converter, usually referred to as a 'charge amplifier'. This uses a high-gain voltage amplifier arranged so that its input capacitance is very high.

Consider the amplifier A, capacitor C_i, and piezo-electric charge generator shown in fig. 4.15(a). Let C_i be the effective input capacitance due to the crystal, cable, and amplifier and let Q be the charge generated by the piezo-electric transducer.

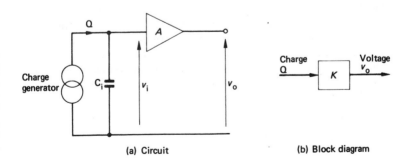

(a) Circuit

(b) Block diagram

Fig. 4.15 The charge amplifier

65

$$v_o = Av_i$$

$$= A\frac{Q}{C_i}$$

where Q = charge in coulombs

and C_i = capacitance in farads

If A and C_i are constants,

$$v_o = KQ \qquad\qquad 4.13$$

i.e. the output voltage v_o is directly proportional to the input charge Q. This may be represented by the block diagram shown in fig. 4.15(b).

The main features of the charge amplifier are

a) the very high input resistance, typically 10^{14} ohms, which reduces the problem of charge leakage from the piezo-electric crystal;

b) the very high input capacitance, typically $1\,\mu F$, which completely swamps the effect of the cable capacitance (typically $10 \times 10^{-3}\mu F$).

Example 4.11 The following details are extracted from the specification of a charge amplifier:

Output voltage ± 10 volts
Input impedance   greater than 10^{14} ohms
Output impedance $0.1\,k\Omega$
Frequency response   d.c. to $150\,kHz$
Noise (r.m.s.) $2\,mV$ maximum [this is the apparent output signal when the input signal is zero]
Linearity error 0.1%
Temperature drift $\pm 0.1\,mV/^\circ C$

a) If the charge-amplifier output voltage is measured by a voltmeter having a resistance of $10\,k\Omega$, determine the reduction in voltage due to the loading effect.

b) What would be the effect of electrically connecting an oscilloscope across the input terminals in an attempt to view the input signal? Assume the oscilloscope has an input impedance of $2\,M\Omega$.

c) When would the specified noise present a problem?

d) Determine the output voltage change if the surrounding temperature changes by $+15^\circ C$. Would this seriously affect the measurements?

e) Could the amplifier be used at an input signal frequency of $180\,kHz$?

a) Let   unloaded voltage = V volts

When loaded,

$$\text{output voltage} = V \times \frac{10\,k\Omega}{10\,k\Omega + 0.1\,k\Omega} = V \times 0.99$$

$$\therefore \qquad \text{reduction} = V(1 - 0.99)$$
$$= 0.01\,V$$

i.e. there is a 1% reduction in output voltage.

b) The oscilloscope input impedance 'shunts' the input impedance of the charge amplifier. The oscilloscope input impedance of $2\,M\Omega$ would be in parallel with the input impedance of the charge amplifier, so their combined impedance presented to the transducer would be less than $2\,M\Omega$. Hence the transducer charge rapidly leaks away.

c) When using the system with very low signal levels.

d) With a drift of $0.1\,mV/^\circ C$, $15^\circ C$ temperature change will give $1.5\,mV$ output voltage due to temperature. This is unlikely seriously to affect the measurement accuracy, since the signal output voltage is likely to be several volts.

e) Yes, but there would be a reduction in gain.

4.5.3 Modulation techniques

The transduced signal may be superimposed on a high-frequency waveform known as the *carrier* in such a way that the original signal may be recovered and displayed. The carrier waveform is thus said to be *modulated* by the signal, and the process of signal recovery is referred to as *demodulation*.

Several modulation techniques are available, among them being

a) amplitude modulation (AM), where the amplitude of the carrier is varied by the transduced signal;

b) frequency or phase modulation (FM), where the instantaneous frequency or phase of the carrier is varied by the signal.

a) Amplitude-modulating systems

In an amplitude-modulating system, the physical quantity being measured is used to vary the amplitude of a high-frequency signal. The resulting varying-amplitude high-frequency signal may then be amplified by a simple inexpensive a.c. amplifier. Figure 4.16 shows an example of an amplitude-modulating system, which is made up of the following devices:

i) a *signal generator* which provides a high-frequency carrier-signal wave, usually of sinusoidal form, for

ii) a *transducer* which senses the quantity being measured and uses this to modulate the carrier signal. A typical example employing this technique is an a.c.-excited resistance-bridge circuit where the resistance changes are converted into bridge output-voltage changes. The modulated high-frequency output voltage from the transducer may then be amplified by

iii) an *a.c. amplifier* which amplifies the modulated waveform. It can be shown that if the frequency of the physical quantity being measured

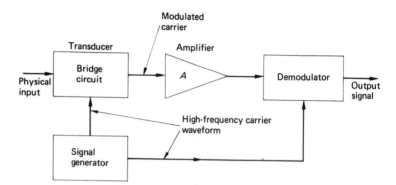

Fig. 4.16 An amplitude-modulating system

varies between 0 and f_1 Hz and the carrier frequency is f_2 then the information input frequency to the amplifier will be between $(f_2 - f_1)$ Hz and $(f_2 + f_1)$ Hz. A typical value for the carrier frequency would be 10 kHz and the upper frequency of the measured quantity would be 20 Hz; thus the information frequencies would lie between 9980 Hz and 10020 Hz.

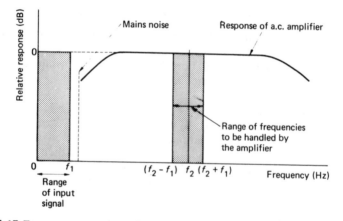

Fig. 4.17 Frequency response of an amplitude-modulating system

The frequency response of a typical a.c. amplifier is shown in fig. 4.17, together with the signal spectrum and the noise due to induced signals from the electrical mains supply. Since the information frequencies are within the flat part of the amplifier's frequency response, they will be amplified; but the noise, not being amplified, is rejected. Following amplification, the original signal is recoverable by

iv) a *demodulator* which recovers the original signal from the modulated waveform. The demodulator has two inputs, one being the modulated carrier waveform, the other a reference signal from the signal generator. When these two signals are in phase, the demodulator gives a positive output; when the signals are in anti-phase the output is negative. This phase-sensitive feature of the demodulator makes it a particularly useful device in measurement systems where the sense as well as the magnitude of the input signal is to be measured. The waveform relationships shown in fig. 4.18 illustrate the behaviour of this modulating and demodulating process.

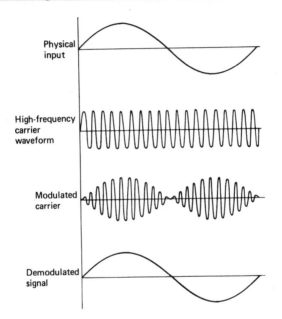

Fig. 4.18 Waveforms in an AM system

b) Frequency-modulating systems
In a frequency-modulating system, the physical quantity being measured is used to vary the instantaneous frequency of a very-high-frequency carrier signal. An example of a frequency-modulating system is shown in fig. 4.19. The transducer, usually a capacitive or inductive type, receives the input signal which it converts into a change of capacitance C or inductance L. These variations in C or L are used to change the resonant frequency of an electronic oscillator which may be thought of as an amplifier employing positive feedback to maintain an unstable condition so that oscillations occur. The frequency changes are sensed by a frequency discriminator which gives a d.c. output voltage proportional to the frequency deviation from a centre or reference frequency.

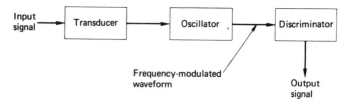

Fig. 4.19 A frequency-modulating system

With zero input physical quantity the discriminator is tuned to give zero output voltage. As the input quantity changes, the oscillator frequency changes giving a variation in the output voltage. In practice, the discriminator is arranged to operate over the linear part of the characteristic shown in fig. 4.20. An examination of this curve will show that the output will be positive for increasing frequency and negative for decreasing frequency. The relationship between the waveforms of an FM system is shown in fig. 4.21.

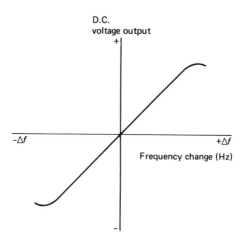

Fig. 4.20 Frequency-discriminator characteristic

Comparison of AM and FM
Any unwanted noise introduced into an AM system will normally produce an unwanted variation in the amplitude of the signal output. Since FM systems are insensitive to amplitude variations, such noise signals which produce amplitude change will not appear at the FM output. In addition, since the AM system's carrier frequency must be at least ten times that of the quantity being measured in order to avoid distortion effects, there is an upper frequency limit to the operational measurement of an AM system.

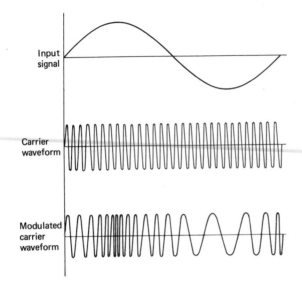

Fig. 4.21 Waveform relationships in an FM system

Although the FM system gives a much better signal-to-noise ratio than AM under similar operating conditions, FM systems are usually more sophisticated and expensive than AM systems.

4.5.4 Bridge circuits
Electrical bridge circuits are used extensively in industrial instrumentation. Although they may be used with resistors, capacitors, or inductors, we will confine our study to the analysis and application details of pure resistance bridges.

a) Null-balance and deflection methods
Consider the resistance Wheatstone-bridge circuit shown in fig. 4.22. For zero output voltage, referred to as the 'balanced condition', the following relationships exist.

$$v_{AB} = v_{AD} \quad \text{i.e.} \quad i_1 R_1 = i_2 R_4$$

and $\quad v_{BC} = v_{DC} \quad \text{i.e.} \quad i_1 R_2 = i_2 R_3$

hence $\quad \dfrac{i_1 R_1}{i_1 R_2} = \dfrac{i_2 R_4}{i_2 R_3}$

$\therefore \qquad \dfrac{R_1}{R_2} = \dfrac{R_4}{R_3}$ \hfill 4.14

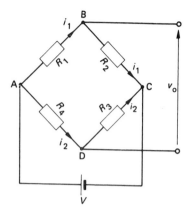

Fig. 4.22 A Wheatstone-bridge circuit

This condition for balance is used in 'null-balance' bridges where one resistor is variable and has an indicator dial calibrated in the applicable engineering units or variables such as strain or temperature which produce the resistance changes of the sensing element. The scale calibration of the output-voltage measuring device, possibly a meter, is unimportant since it is used only as a null or zero-voltage indicator.

The value of the calibrated resistor is adjusted to give zero output voltage and the calibrated dial gives an indication of the resistance changes required to balance the system. When the bridge is used so that its output voltage v_0 is used to give an indication of the resistance changes, the bridge is then referred to as a *deflection bridge* since its output voltage may be used to produce a deflection of the pointer on an electrical meter.

b) Wheatstone-bridge circuit analysis
Consider the Wheatstone bridge illustrated in fig. 4.22, made up of four resistors and excited with a voltage V. The output voltage from the bridge will change as the bridge becomes unbalanced. Inserting the following initial conditions simplifies the analysis.

Assume the bridge is *initially balanced* and let

$$R_1 = R_2 = R_3 = R_4 = R$$

then $\qquad v_{AD} = v_{AB} = \dfrac{V}{2}$

and $\qquad v_0 = v_{AB} - v_{AD} = 0$

Let R_1 change by an amount ΔR_1 to $(R_1 + \Delta R_1)$ so that v_{AB} will change.

New value of $v_{AB} = \dfrac{R_1 + \Delta R_1}{R_1 + \Delta R_1 + R_2} \times V$

but if $R_1 = R_2 = R$

then
$$v_{AB} = \frac{R + \Delta R}{2R + \Delta R} \times V$$

$$\therefore \quad v_o = v_{AB} - v_{AD} = V \left(\frac{R + \Delta R}{2R + \Delta R} - \frac{1}{2} \right)$$

$$= \frac{\Delta R}{4R + 2\Delta R} \times V$$

and if $\Delta R \ll R$

$$v_o = \frac{V}{4} \frac{\Delta R}{R} \qquad\qquad 4.15$$

By a similar analysis it can also be shown that, if two resistors R_1 and R_2 vary, the expression becomes

$$v_o = \frac{V}{4} \left(\frac{\Delta R_1}{R} - \frac{\Delta R_2}{R} \right) \qquad\qquad 4.16$$

Thus if the resistance changes are of the same sign and magnitude they cancel each other out and the output voltage is zero. This effect is put to good use in strain-gauge temperature compensation, which is discussed in chapter 8.

If the resistance changes are of opposite sign and equal in magnitude, the output voltage v_o becomes

$$v_o = \frac{V}{2} \times \frac{\Delta R}{R}$$

Similarly, if the changes are to $R_1 + \Delta R_1$, $R_2 - \Delta R_2$, $R_3 + \Delta R_3$, and $R_4 - \Delta R_4$ as used in a full active bridge circuit where all the resistors vary in value, the corresponding output voltage is given by,

$$v_o = V \frac{\Delta R}{R} \qquad\qquad 4.17$$

where $\Delta R = \Delta R_1 = \Delta R_2 = \Delta R_3 = \Delta R_4$

Example 4.12 A resistance Wheatstone-bridge circuit made up of four resistors each of value $120\,\Omega$ has an excitation voltage of $5\,V$. Determine the output voltage change when one resistor's value changes by $1.2\,\Omega$.

Using equation 4.15,

$$v_o = \frac{V}{4} \frac{\Delta R}{R}$$

$$= \frac{5\,V}{4} \times \frac{1.2\,\Omega}{120\,\Omega} = 12.5\,mV$$

Example 4.13 A symmetrical Wheatstone bridge is made up of four equal-value resistors. When all four resistors change in value by 1%, the sense of the changes being such that maximum sensitivity is obtained, the output voltage alters by 100mV. Determine the value of the excitation voltage.

Using equation 4.17,

$$v_o = V \frac{\Delta R}{R}$$

$$V = \frac{v_o R}{\Delta R}$$

$$= 100 \times 10^{-3} \text{V} \times \frac{100}{1} = 10 \text{V}$$

c) A.C. and d.c. excitation of bridges

The bridge circuit may be excited by either a d.c. voltage or a high-frequency sinusoidal carrier signal. The latter is a particular application of amplitude modulation where the bridge output voltage is modulated by the input signal. The relative merits of the two techniques are indicated in Table 4.1.

Table 4.1 Relative merits of a.c. and d.c. excitation

D.C.	A.C.
D.C. amplifiers, if required, are more expensive and experience drift problems.	A.C. amplifiers are simple to construct and free from drift.
Suitable for high-frequency measurement, e.g. of dynamic strain.	Upper frequency limit of measurement restricted to one tenth of the carrier frequency. Most suitable for static strain measurement.
Balancing and zero adjustment simple.	Balancing requires reactive and resistive balancing to overcome phase differences caused by self-inductances and capacitances of the gauges.
Suffers from unwanted 'noise' signals caused by electromagnetically induced signals, thermoelectric effects, and drift.	Free from noise caused by electromagnetically induced signals, thermoelectric effects, and drift.

Exercises on chapter 4

1 A compound lever comprises two levers A and B. If lever A has a length magnifying ratio of 4.2:1 and the displacement gain of the device is 18, determine the length magnifying ratio of lever B. [4.29:1]

2 Explain the differences between a.c. and d.c. amplifiers. Illustrate your answer with a frequency-response graph.

3 If the voltage gain of an electronic amplifier is increased 6 times, express this increase in decibels. [15.56dB]

4 A frequency-response test on an amplifier yielded the following results:

Frequency (Hz)	10	20	40	80	160	320	640	1280	2560	5120
Voltage output (V)	2	7	12.3	16	17	17.8	18	18.1	17.2	10

Draw the frequency-response graph of relative response in dB versus frequency, using the voltage at 640 Hz as your reference.

5 State five effects of negative feedback when applied to an amplifier.

6 An amplifier has a gain without feedback of 10^4. If 5% of the output voltage is fed back in anti-phase to the input voltage, determine the resulting gain with feedback. [19.96]

7 State three properties of an operational amplifier and discuss the implications of two of these properties.

8 The summer amplifier shown in fig. 4.11 has the following details: $R_1 = 2\,\text{k}\Omega$, $R_2 = 3\,\text{k}\Omega$, $R_3 = 1.5\,\text{k}\Omega$, $R_f = 12\,\text{k}\Omega$, $v_1 = +2\,\text{V}$, $v_2 = +3\,\text{V}$, $v_3 = -2\,\text{V}$. Calculate the output voltage v_o. [$-8\,\text{V}$]

9 What is a charge amplifier? State two features of such an amplifier.

10 State the condition which must exist for 'balance' in a Wheatstone-bridge arrangement of resistors.

11 A symmetrical bridge circuit is excited by 10V d.c. Assuming the bridge is initially balanced, determine the output voltage of the bridge if the value of one resistor changes by 1%. [0.025V]

12 If in the Wheatstone-bridge circuit shown in fig. 4.22 the resistor values change from the balance condition to the following: $(R_1 + \Delta R_1)$, $(R_2 - \Delta R_2)$, $(R_3 - \Delta R_3)$, $(R_4 + \Delta R_4)$, determine the expression for the output voltage in terms of the excitation voltage and resistance values.

13 Explain the differences between the amplitude- and frequency-modulation techniques used in measuring systems.

5   Recording and display equipment

5.1 Introduction

The recorder or display unit is the last element in the measuring system and is the component that provides the results of the measurement. Its selection is therefore just as critical as that of the correct transducer or signal conditioner – after providing a signal which was a faithful reproduction of the measurand, it would be foolish to introduce errors by using an unsuitable recorder.

The difference between a recorder and a display unit is that the recorder produces a permanent record of the signal while the display unit does not. Examples of recorders are u.v. recorders, pen recorders, X–Y plotters, and tape recorders. The speedometer in a car, the mercury level and scale in a thermometer, and the trace of a normal oscilloscope are all examples of displays.

In a rapidly developing field, there exists a large range of sophisticated recording equipment which uses microprocessors and semiconductor memories to record transient signals. These, however, are beyond the scope of the book, and the remainder of this chapter will concentrate on the more common types of recorder and display units.

5.2 Mechanical pointers

It is possible to connect a pointer and scale to a number of transducing elements to obtain a display of the parameter being measured. The Bourdon-tube pressure gauge is one example, where the movement of the tube is transmitted through levers and gearing to the pointer on the gauge face. Another example is illustrated in fig. 5.1, which shows a temperature-measuring device which uses a bimetallic strip wound in the form of a helix. As the temperature rises, the coil unwinds (due to the different coefficients of expansion of the two metals) and the pointer rotates.

Once a pointer movement which is proportional to the measurand has been obtained, it is a relatively simple matter to attach a light pen in order to provide a permanent record. Paper moving past the pen at a constant speed would then give a plot of the variable against time. One common example of this type of recorder is the pressure-chart recorder which is illustrated in fig. 5.2. In this case the power to drive the marking system is obtained directly from the pressure signal, and this is called a direct-recording instrument. It is relatively cheap and robust but requires the chart to be changed after one revolution, otherwise the two super-

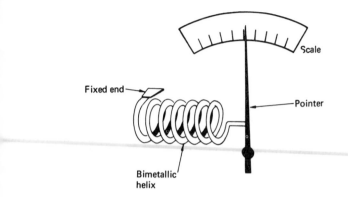

Fig. 5.1 A simple temperature-measuring device

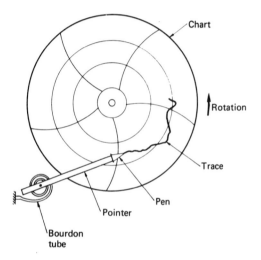

Fig. 5.2 A pressure-chart recorder

imposed traces could be confused. A typical paper speed would be one revolution in twenty-four hours.

Other examples of direct-recording instruments are the barograph to measure atmospheric pressure and the seismograph to monitor the earth's vibrations.

Because of the inertia of the mechanical system, recorders of this type are restricted to recording slowly changing signals over long periods of time. Response times of several seconds are typical.

For the remainder of this chapter, electrical recorders and display units will be considered. These offer a number of advantages over mechanical devices, including greater flexibility in handling a wide range of signals and better frequency-response characteristics.

5.3 The moving-coil mechanism

The mechanism illustrated in fig. 5.3 is known as the d'Arsonval movement and forms the basis of a number of electrical recorders and displays. A current of the order of a few milliamperes flows in the coil and produces an electromagnetic field which tries to align itself with the field of the permanent magnets. The electromagnetic torque causes a rotation of the coil, which is opposed by the torsion of the spiral springs. When the spring torque and the electromagnetic torque are balanced, the rotation ceases and an angular deflection of the mechanism is produced.

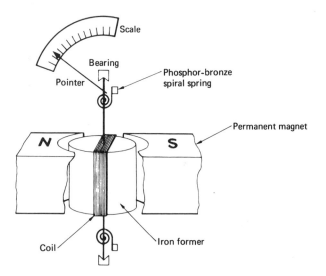

Fig. 5.3 The basic moving-coil mechanism

By using springs with linear torque characteristics and correctly shaped permanent magnets, it is possible to obtain a linear relationship between the coil current and the angular displacement of the pointer over a range of approximately 60°.

This mechanism is found in moving-coil ammeters, voltmeters, and multimeters and in display units in control panels.

5.4 Pen recorders

The moving-coil mechanism is used in a number of commercially available pen recorders. By attaching a pen to the pointer and driving paper past it at a constant speed, a permanent record is produced. Other types of recorder use the same basic principle with variations such as

a) a heated stylus which passes over heat-sensitive paper;
b) a high voltage on a pointer which burns a trace into special carbon-layered paper;
c) a mirror attached to the moving-coil mechanism which deflects ultra-violet light on to light-sensitive paper. This type will be dealt with in greater detail in the next section.

A range of input signals can be recorded with these pen recorders, in the same way as a moving-coil multimeter can display a wide range of voltages or currents. A typical range of sensitivities is 10mm/V to 0.05mm/V.

One disadvantage is that movements are restricted to small angles, typically ±15°, due to the difficulties in converting the angular rotation of the pointer to a displacement across the chart. The use of chart paper with curved grid lines is common, but familiar shapes such as sine waves become distorted. A knife-edge used with a broad stylus eliminates the need for curved grid lines but increases the non-linearity of the movement to approximately ±2% deviation for the ±15° movement. This system is illustrated in fig. 5.4.

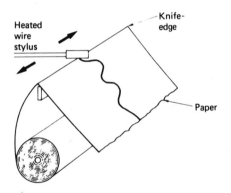

Heated wire stylus

Knife-edge

Paper

Fig. 5.4 The arrangement of a recorder with a knife-edge and a broad stylus

5.5 Ultra-violet recorders

The basic construction of the ultra-violet (u.v.) recorder is shown in fig. 5.5. Current in the moving-coil galvanometer causes rotation of the mirror which deflects the light beam on to the sensitive paper. The use of the mirror system increases the optical arm length, thus giving a greater movement of the trace across the paper for a given angular displacement of the galvanometer mirror. Optical arm lengths of 350mm are typical, although the overall length from the galvanometer to the paper is much less than this. Although the u.v. recorder has been largely superseded, it is still in common use.

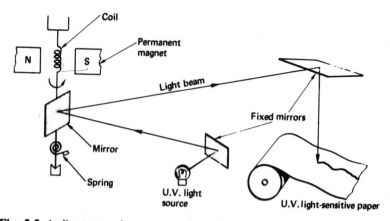

Fig. 5.5 A diagrammatic representation of a u.v. recorder

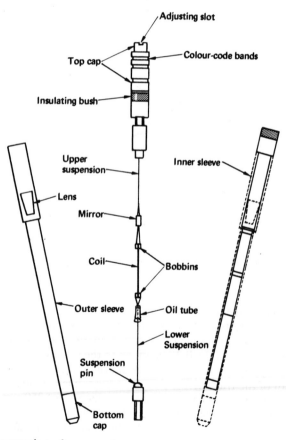

Fig. 5.6 Construction of a u.v. galvonometer

The galvanometer construction is illustrated in fig. 5.6, which shows that the units are removable and interchangeable. They are mounted in a magnetic block in rows of up to 25 galvanometers (see fig. 5.7). Apart from the paper drive, the galvanometer mirror-and-coil assembly is the only moving mechanism and therefore determines the dynamic performance of the u.v. recorder.

5.5.1 Typical specification
Number of channels   6, 12, or 25
Writing speed   greater than 1500 m/s, depending on conditions and materials
Maximum spot deflection/paper width   203 mm
Range of paper speeds   5 m/s to 1 mm/min, within ±1%
Light source   50 W high-pressure mercury vapour lamp
Timing-line intervals   0.002 s, 0.01 s, 0.1 s, 1 s, 10 s, or minutes
Record duration   manual, 0.5 s, 1 s, 2.5 s, 10 s, 20 s

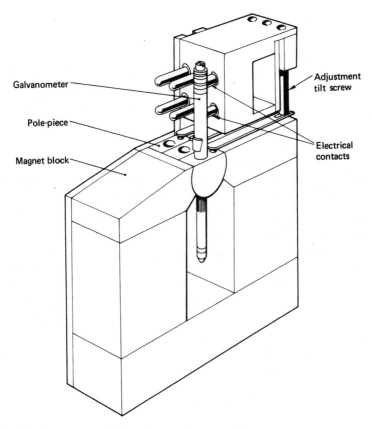

Fig. 5.7 Arrangement of a galvanometer assembly

5.5.2 Galvanometer sensitivity

Galvanometer sensitivity is defined as the galvanometer spot deflection produced by unit current. Manufacturers often quote values which are the inverse of those given by this definition.

Typical values are

20 Hz galvanometer   2 cm/mA (0.5 mA/cm)
10 kHz galvanometer   0.043 cm/mA (23 mA/cm)

The sensitivity values enable the current required for a given size of trace to be determined.

Example 5.1   A 20 Hz galvanometer is required to give a full-paper-width deflection of 200 mm. Determine the current needed.

$$\text{Deflection} = 20\,\text{cm}$$

$$\text{Sensitivity} = 2\,\text{cm}/\mu\text{A}$$

$$\therefore \quad \text{required current} = \frac{20\,\text{cm}}{2\,\text{cm}/\mu\text{A}} = 10\,\mu\text{A}$$

Table 5.1 shows the specifications of a range of galvanometers, and it should be noted that the higher the natural frequency the lower the value of galvanometer sensitivity. This is due to the fact that, in order to increase the natural frequency, stiffer springs are used and therefore more current is required to produce a given deflection, i.e. the sensitivity is reduced.

5.5.3 Galvanometer dynamics

Since the galvanometer possesses mass, damping, and spring stiffness, it is a second-order system with the standard step and frequency responses dealt with in chapter 2. It was stated there that the optimum condition for a second-order measuring system occurs when the damping ratio is 0.64, and with the u.v. galvanometer this is achieved in one of two ways.

Low-frequency galvanometers use electromagnetic damping, where an induced current opposes the motion. In order to achieve the correct amount of damping, a definite value of resistance must be connected across the galvanometer terminals. This is called the damping resistance R_d and is typically 250 Ω. In some circuits R_d may not be one single resistance but the equivalent resistance of a number of resistors.

For higher-frequency galvanometers fluid damping is employed, where the whole assembly is immersed in a silicone damping fluid to give the optimum condition.

5.5.4 Galvanometer-matching networks

The functions of a matching network are

a) to prevent the galvanometer taking excessive current,

Table 5.1 Typical u.v. galvanometer specifications

Type	Natural frequency f_n (Hz)	Galvanometer resistance R_g (ohms)	Damping resistance R_d (ohms)	Sensitivity			Maximum current (mA) (5s maximum)
				cm/mA	mA/cm*	mV/cm*	
1	20	45	250	2000	0.0005	0.021	5
2	40	40	250	769.2	0.0013	0.052	20
3	100	57	250	476.2	0.0021	0.120	20
4	200	80	250	100.0	0.010	0.80	20
5	500	120	250	20.0	0.050	6.0	20
6	1000	32	Fluid	2.0	0.50	16.0	75
7	2200	32	Fluid	0.512	1.95	62.3	75
8	5000	32	Fluid	0.10	10.0	320	75
9	10000	32	Fluid	0.0435	23.0	736	75
10	20000	85	Fluid	0.0171	58.0	4930	75

* Sensitivities quoted in manufacturers' specifications.

b) to provide the correct value of damping resistance,
c) to avoid overloading the signal source.

In the matching circuits shown in fig. 5.8, R_g is the galvanometer resistance, R_d is the required damping resistance, R_s is the source resistance, v_s is source voltage, and i_g is the galvo current.

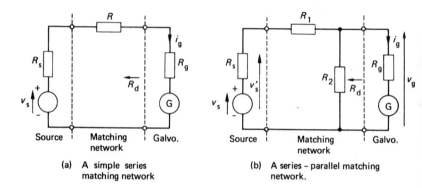

(a) A simple series matching network

(b) A series – parallel matching network.

Fig. 5.8 Matching networks

Circuit (a) shows a low-voltage source or a fluid-damped galvanometer with a series connection through resistor R. R_d is the resistance seen from the galvanometer terminals and is therefore the series combination of R and R_s,

i.e. $R_d = R + R_s$

and galvo current $i_g = \dfrac{v_s}{R_s + R + R_g}$

Circuit (b) is perhaps a more practical situation, where the source voltage is too high for direct connection and some attenuation or reduction in signal is required. If the series connection as in circuit (a) were used, the resistance would be too high to satisfy the damping requirement. R_2 is placed across the galvanometer terminals and R_d is the parallel combination of R_2 and $R_1 + R_s$,

$\therefore \quad R_d = \dfrac{R_2 (R_1 + R_s)}{R_2 + (R_1 + R_s)}$

Attenuation of the network is given by

$\dfrac{v_g}{v_s'} = \dfrac{R_2 \parallel R_g}{R_1 + R_2 \parallel R_g}$

where $R_2 \parallel R_g$ means the parallel combination of R_2 and R_g, i.e. $R_2 \parallel R_g = \dfrac{R_2 R_g}{R_2 + R_g}$

and v_s' is the source terminal voltage.

$\therefore \quad \dfrac{v_g}{v_s'} = \dfrac{R_2 R_g}{R_1(R_2 + R_g) + R_2 R_g}$

Example 5.2 The 1000 Hz fluid-damped u.v. galvanometer specified in Table 5.1 is to be used to record vibrations on an aircraft engine test. If the signal from the measuring equipment is 2 V peak-to-peak, design a matching circuit to obtain a trace width of 15 cm.

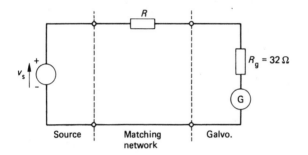

Fig. 5.9 Matching network for example 5.2

Figure 5.9 shows a suitable matching network.
  From 1000 Hz specification,

  galvanometer sensitivity = 2.0 cm/mA

therefore, for a deflection of 15 cm,

  current required $= \dfrac{15\,\text{cm}}{2.0\,\text{cm/mA}} = 7.5\,\text{mA}$

$\therefore$ total resistance required $= \dfrac{2\,\text{V}}{7.5 \times 10^{-3}\,\text{A}}$

$= 267\,\Omega$

$\therefore$ series resistance (R) required $= 267\,\Omega - 32\,\Omega$

$= 234\,\Omega$

Example 5.3 A charge amplifier produces a peak-to-peak voltage of 1.5 V from a piezo-electric pressure transducer measuring pressure fluctuations in a compressor. The 200 Hz galvanometer specified in Table

5.1 is to be used to obtain a u.v. record. Design a suitable matching network if the trace width is to be 100 mm. Neglect signal-source resistance.

From Table 5.1

$$R_g = 80\,\Omega \quad \text{and} \quad R_d = 250\,\Omega$$

Sensitivity $= 100\,\text{cm/mA} \equiv 0.01\,\text{mA/cm}$ or $0.8\,\text{mV/cm}$

For a trace of 100 mm,

voltage required $= 0.8\,\text{mV/cm} \times 10\,\text{cm} = 8\,\text{mV}$

A matching circuit of the type shown in fig. 5.8(b) is therefore required.

Attenuation required $= \dfrac{8 \times 10^{-3}\,\text{V}}{1.5\,\text{V}} = 5.33 \times 10^{-3}$

$$\therefore \quad \frac{R_2 \times 80\,\Omega}{R_1 (R_2 + 80\,\Omega) + (R_2 \times 80\,\Omega)} = 5.33 \times 10^{-3} \qquad \text{(i)}$$

Also $\quad R_d = \dfrac{R_1 R_2}{R_1 + R_2}$

$$\therefore \quad 250\,\Omega = \frac{R_1 R_2}{R_1 + R_2}$$

If $\quad R_1 \gg R_2 \quad$ then $\quad R_d \simeq \dfrac{R_1 R_2}{R_1} = R_2$

$\therefore \quad$ make $R_2 = 250\,\Omega$

Substituting this value in equation (i) gives

$$\frac{250\,\Omega \times 80\,\Omega}{(R_1 \times 330\,\Omega) + (250\,\Omega \times 80\,\Omega)} = 5.33 \times 10^{-3}$$

Transposing,

$$R_1 \times 330\,\Omega + 250\,\Omega \times 80\,\Omega = \frac{250\,\Omega \times 80\,\Omega}{5.33 \times 10^{-3}}$$

$\therefore \quad R_1 \times 330\,\Omega = (3.75 \times 10^6 - 20000)\,\Omega^2$

$\qquad \qquad \simeq 3.75 \times 10^6\,\Omega^2 \quad$ since $20000 \ll 3.75 \times 10^6$

$$\therefore \quad R_1 = \frac{3.75 \times 10^6\,\Omega^2}{330\,\Omega}$$

$$= 11364\,\Omega \simeq 11.4\,\text{k}\Omega$$

The required matching circuit is shown in fig. 5.10. Alternatively, to give some adjustment of the size of the trace, a potentiometer could be included as illustrated in fig. 5.11. Note that resistance values are only

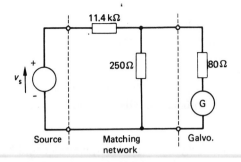

Fig. 5.10 Matching network for example 5.3

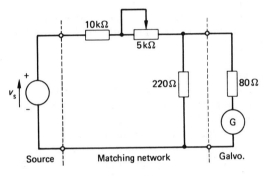

Fig. 5.11 A practical matching network for example 5.3

nominal and can vary by ±10% and also that 250Ω is a non-standard value so that in practice 220Ω would be used.

5.5.5 Selection of u.v. galvanometers
The main criterion for the selection of a u.v. galvanometer is the frequency response, for this will determine whether or not the instrument will follow a changing signal to within the desired measurement accuracy. If the damping ratio is 0.64 then the galvanometer will respond to sine waves within ±3% up to frequencies of 60% of the galvanometer natural frequency. Thus a prior knowledge of the frequency of the signal is required and is obtained by viewing on an oscilloscope.

Example 5.4 The signal frequency from a strain measurement on a vibrating body was found to be 239 Hz. From Table 5.1, select a suitable galvanometer to record the signal.

To be within ±3% error,

$$\frac{\text{required galvanometer}}{\text{natural frequency}} = \frac{239\,\text{Hz}}{0.6}$$

$$= 398\,\text{Hz or higher}$$

Therefore type 5, having $f_n = 500\,\text{Hz}$, would be used.

Galvanometers with higher natural frequencies could be used but they require greater currents for a given trace size and the demand on the signal source could be excessive. If harmonics (i.e. integer multiples of the fundamental frequency) occur in the waveform of the signal then again the higher-frequency galvanometer could be used to obtain a more faithful recording. The presence of the third harmonic in example 5.4 would require the galvanometer to respond to a frequency of three times the fundamental, namely 1200 Hz, and so the 2200 Hz galvanometer would be selected.

5.5.6 Modern developments in u.v. recording

Since the inertia of the galvanometer is the limiting factor in the frequency response of the u.v. recorder, ways of removing it from the system have been investigated. One system has been developed which uses an array of light gates which have to be energised before they allow polarised light to pass through them. As the magnitude of the input signal varies, so different gates across the array are energised to give the desired trace. Since no moving parts are involved, the step response shows no overshoot. The frequency response is limited by the speed at which the gates can be switched on, but a typical response is

d.c. to 5 kHz for sine waves
d.c. to 10 kHz for square waves

Another system uses a cathode-ray tube with a fibre-optic stack which causes the light beams to be emitted at right angles to the oscilloscope screen. Focusing systems are eliminated, and u.v. paper can be placed on the screen to produce X–Y plots. Alternatively, the paper can be driven past the screen at a constant speed to produce a normal Y–t trace.

5.6 Servo-recorders

5.6.1 Closed-loop servo-recorder and display

A servo-recorder is one in which an external power supply is needed to move the pen. In general terms, a closed-loop control system or servo-system can be represented by the block diagram shown in fig. 5.12. It can be seen that the output is fed back in a negative sense, in order to create an error or difference signal. The controller senses the error and causes

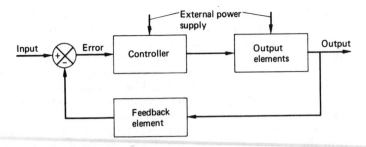

Fig. 5.12 Block diagram of a closed-loop control system

an appropriate change of output to reduce the error to zero, at which point the output and input will be equal. Closed-loop control systems will be considered in greater detail in chapter 13.

In the case of the pen recorder, the input is usually in the form of a voltage and the output is the pen position, as illustrated in fig. 5.13. The feedback element is the potentiometer, which provides a voltage which is proportional to the pen position. The error signal is a voltage difference and this is amplified to drive the d.c. motor which, through gearing and pulleys, moves the pen.

These recorders are sometimes called potentiometric or self-balancing recorders and are commonly used to record temperatures from thermocouple circuits.

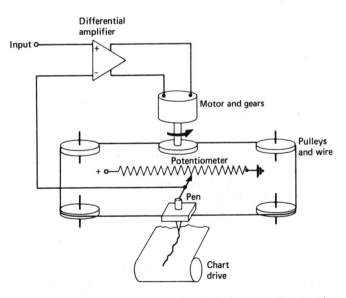

Fig. 5.13 Diagrammatic arrangement of a potentiometer pen recorder

The use of feedback in this type of recorder makes the accuracy and linearity of the system more dependent on the potentiometer than on the motor-and-drive system. Since the potentiometer specification can be met with closer tolerances than those of the motor and pulley, the system is more accurate than an open-loop type. The potentiometer is the most important part of the recorder and should be maintained properly and regularly, as dirt on the wiper contact can cause judder on the recorder. Another problem that could arise is dead-band, which in this case is the error required to produce a signal which is large enough to overcome friction in the system. Obviously the dead-band should be as low as possible, to reduce errors.

Closed-loop displays are also available, and fig. 5.14 shows a strip indicator which uses a feedback system to control the strip height.

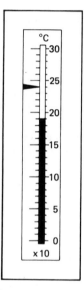

Fig. 5.14 A potentiometer strip indicator

5.6.2 *Typical closed-loop recorder specification* (illustrated in fig. 5.15)
Range of chart speeds   5 mm/h to 3600 mm/h
Intrinsic error   ±0.5% span maximum
Dead-band   ±0.3% span maximum
Minimum span   5 mV
Response time   1 s

5.6.3 X–Y plotters
The X–Y plotter is an extension of the closed-loop servo-recorder, having a pen that can be positioned in two axes with the paper remaining stationary. A typical X–Y plotter is shown in fig. 5.16.

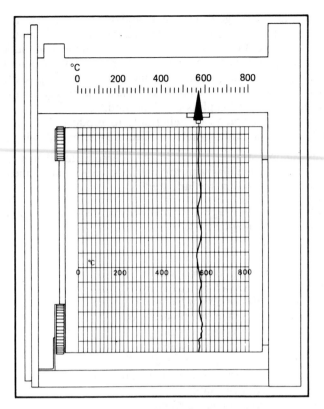

Fig. 5.15 A diagram of the Kent Clearspan P120L recorder

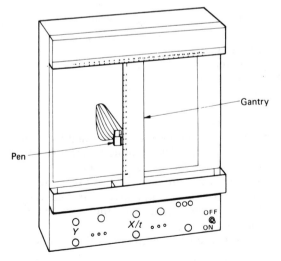

Fig. 5.16 An X–Y plotter

The pen and carriage can move up and down the gantry in the Y-axis, and this position is controlled automatically by means of a motor, pulleys, and a linear potentiometer. The gantry itself can move in the X-axis and its position is controlled in a similar manner to that of the pen. A range of voltages applied to the X and Y channels can therefore give plots of one variable against another. Time-base units are normally incorporated in X–Y plotters to give Y–t plots where desired.

This major advantage of being able to produce X–Y plots of one variable against another variable is illustrated in the following examples:

a) stress against strain for a specimen undergoing tensile testing,
b) pressure–volume diagrams for internal-combustion engines,
c) pressure against flow for lung studies.

The major disadvantage is the relatively poor speed of response, due to the inertia of the moving parts.

5.6.4 Typical X–Y plotter specification
Paper size   DIN A3 (42 cm × 29.7 cm)
Plotting area   38 cm × 28 cm (X × Y)
Paper hold down   (a) magnetic strips, or
                             (b) vacuum, or
                             (c) electrostatic
Paper location   illuminated graticules
Pen   capillary with reservoir (rechargeable), or
          fibre-tipped (disposable)
Maximum velocity, Y-axis   200 cm/s
Maximum velocity, X-axis   150 cm/s
Slewing speed   250 cm/s
Time to go from 0 to 100% of full scale in Y-axis (28 cm)   less than 160 ms
Time to go from 0 to 100% of full scale in X-axis (38 cm)   less than 300 ms
Linearity   better than 0.1% f.s.d.
Repeatability   better than 0.1% f.s.d.
Input   100 mV/cm
Input impedance   greater than 10 MΩ

Notes on specifications
When setting up the graph paper, it is important to ensure that the axes are vertical and horizontal. This is done by means of illuminated light strips which shine through the paper at each end.

The maximum velocities differ in the two axes, due to the pen gantry which moves in the X-axis having a greater inertia than the pen which moves in the Y-axis. Response times will therefore be larger in the x-axis, due to lower speeds.

Slewing speed is the vector sum of the writing speeds in each axis;

i.e.   maximum slewing speed $= \sqrt{v^2_{X\text{max.}} + v^2_{Y\text{max.}}}$

where $v_{Xmax.}$ and $v_{Ymax.}$ are the maximum velocities in the X- and Y-axes respectively.

$\therefore$ maximum slewing speed $= \sqrt{200^2 + 150^2}\,\text{cm/s}$

$= 250\,\text{cm/s}$ as in the specification

Finally, the time taken to move from 0 to 100% in the Y-axis is *not* equal to (distance moved in Y-axis) ÷ (maximum Y velocity), since time is required for the servo-system to accelerate to the maximum velocity.

5.6.5 Relationship between signal size and frequency for the X-Y plotter

Consider a sine wave resulting in an amplitude of A cm on the X-Y plotter and having a frequency of f Hz. The equation for the displacement y is

$$y = A\sin 2\pi ft$$

$$\therefore \frac{dy}{dt} = 2\pi fA\cos 2\pi ft$$

Thus it can be seen that the amplitude of the velocity $dy/dt = 2\pi fA$ and therefore this is the maximum value of velocity to be reached. To err on the safe side, assume that the maximum pen velocity is 150 cm/s,

$\therefore$ $2\pi fA = 150$ 5.1

It follows, therefore, that, if the amplitude of the sine wave is to be increased, the frequency which the X-Y plotter can handle will be decreased, since the product of f and A is a constant.

Example 5.5 Determine the maximum allowable frequency if the amplitude of the sine wave is (a) 2 cm and (b) 4 cm. Also determine the maximum allowable amplitude of sine wave if the frequency is 1 Hz.

a) From equation 5.1,

$$f_{max.} = \frac{150\,\text{cm/s}}{2\pi \times 2\,\text{cm}}$$

$$= 11.9\,\text{Hz} \simeq 12\,\text{Hz}$$

b) For a 4 cm amplitude trace,

$$f_{max.} = 6\,\text{Hz}$$

If a frequency of 1 Hz is to be plotted then the maximum allowable amplitude is given by

$$A_{max.} = \frac{150\,\text{cm/s}}{2\pi \times 1\,\text{Hz}} = 24\,\text{cm}$$

The peak-to-peak value of the sine wave would therefore be 48 cm, which is larger than the X-Y plotter dimensions. Obviously the X-Y plotter sensitivity would be reduced in this case to give a trace size which fits the graph paper, but it does mean that the plotter could follow frequencies of 1 Hz or less for any trace size up to A3.

Remember that equation 5.1 applies only to the X–Y plotter specified in section 5.6.4.

5.7 The cathode-ray oscilloscope

The cathode-ray oscilloscope is essentially a voltage-measuring device where the deflection of an electron beam is caused by an input voltage. When the beam of high-velocity electrons strikes the phosphor-coated screen, a spot of light is produced which can be deflected in both the X- and Y-axes. The input signal to be displayed is normally applied to the Y-deflection system, while the X deflection occurs at a constant rate to give a 'time base' to the display. It is, however, possible to apply one input signal to the Y channel and another signal to the X channel to give an X–Y rather than a Y–t display.

Three main factors make the oscilloscope the ideal device to carry out an initial examination of the output from an electrical measuring system. They are

a) the excellent frequency response of typically up to 50 MHz. This is due to the electron having negligible mass and so being able to respond rapidly to changes of deflection.
b) the high input impedance of approximately 1 MΩ. The signal source is therefore not loaded excessively.
c) a visual display of the waveform is obtained, allowing the frequency and maximum values to be determined.

The basic construction of the cathode-ray tube is shown in figs 5.17(a) and (b), where both the electrostatic and the electromagnetic types are

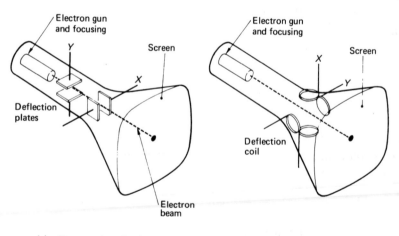

(a) Electrostatic deflection   (b) Electromagnetic deflection

Fig. 5.17 The basic construction of the cathode-ray tube

illustrated. In order to simplify the diagrams, the constructional details of the electron gun and the focusing arrangements have been omitted, for it is in the deflection system that the major differences occur.

In the case of the electrostatic deflection systems, fig. 5.17(a), a deflection occurs in the Y-axis when a voltage is applied to the Y-plates. A positive potential will attract the negatively charged electrons and cause the beam to move towards it, while a negative potential will repel the beam. Typical sensitivities are of the order of 0.02 cm to 0.05 cm deflection per volt.

A magnetic field can also deflect an electron beam, and this is the principle involved in the electromagnetic deflection system, fig. 5.17(b). It should be noted that the Y-deflection coils lie along the X-axis and vice versa, due to the fact that the magnetic field is applied at right angles to the beam to give a deflection which is at right angles to both beam and field. Greater deflections are possible with this type of system, and it is used in television sets to give acceptably large screen sizes.

The two major differences between the two types of oscilloscope are the screen size and the frequency response. The electromagnetic system results in screens of 25 cm × 30 cm, compared with 8 cm × 10 cm for the electrostatic. However, the electrostatic type has a far superior frequency response of 0 to 10 MHz for the basic oscilloscope compared with only 0 to 10 kHz for the electromagnetic type. It is this advantage which results in more electrostatic oscilloscopes being used in measurement systems, and the remainder of this section will be devoted to this type.

Figure 5.18 is a block diagram which illustrates the component parts of a typical cathode-ray oscilloscope. The vertical amplifier is needed to extend the range of input voltages which would give observable deflections; for example, a signal of the order of millivolts would require

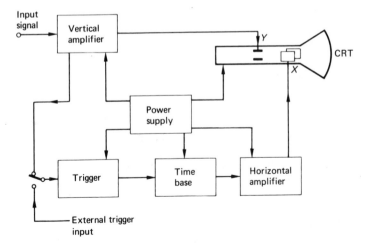

Fig. 5.18 Block diagram of a typical oscilloscope

amplification to give a measurable deflection. In order to obtain a display which is effectively stationary, it is necessary to ensure that the display always starts from the same point on the cycle. The trigger unit performs this function, and a stationary picture can be produced as long as the frequency is not too low and the waveform does not alter.

5.7.1 Typical cathode-ray-oscilloscope specification

Telequipment model D1011 oscilloscope (shown in fig. 5.19)
Vertical system   two input channels
Bandwidth:
   d.c. coupled   d.c. to 10 MHz (-3 dB)
   a.c. coupled   8 Hz to 10 MHz (-3 dB)
Rise time   35 ns
Deflection factors   5 mV/div to 20 V/div
Maximum scan   amplitude 8 divs (6 divs at 10 MHz)
Voltage measurement accuracy $\pm 5\%$
Input impedance   1 MΩ
Input conditions   switched choice of d.c., a.c., or ground. The third position grounds the input of the attenuator but not the signal input.
Maximum input voltage   500 V d.c. or a.c. peak
Operating modes:
   channel 2 only
   channels 1 and 2 chopped or alternated
   $X-Y$ – channel 2 is the vertical input and channel 1 is the horizontal input
Horizontal system:
   sweep speeds   0.2 s/div to 0.2 μs/div $\pm 5\%$
Triggering
   Modes   automatic
             normal
             TV
   Sources   internal channel 2
             external
             line
   Polarity   positive or negative
   Trigger level   variable control selects virtually any point of the positive or negative slope of the input signal.
Cathode-ray tube
   Display area   8 divs × 10 divs (1 div = 1 cm)
   Phosphor   P31
   Accelerating potential   1.8 kV

Notes on specifications
In the chopped mode, the single electron beam is switched rapidly between channels 1 and 2 to give the appearance of two traces. This

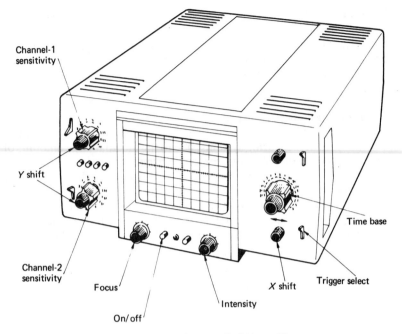

Fig. 5.19 A diagram of the Telequipment D1011 oscilloscope

technique could not be used for higher-frequency signals because unacceptable gaps would appear on the trace.

For higher frequencies, the alternate mode is selected, to give one sweep for channel 1 then one sweep for channel 2, etc.

The triggering modes are

a) automatic – where the trace would be observed as a straight line in the absence of a signal;
b) normal – where the triggering is achieved only with a signal present;
c) TV – where the triggering is achieved by means of signals generated for TV sets. This mode is not normally used in measurement systems.

The triggering sources are either from the signal on channel 2, from some other external signal, or from the mains voltage (line). The requirement for triggering is a minimum of 0.5 div of channel-2 signal or 0.5 V for an external signal. Any signals smaller than these values will not cause the trigger circuit to operate.

5.7.2 The storage oscilloscope

This type of oscilloscope is very useful where a non-repetitive signal is to be displayed, for example a pressure wave moving through a metal bar which has been struck at one end. The trace will be stored for some considerable time, allowing it to be analysed at leisure.

Very briefly, the cathode-ray tube has a mesh inside it near to the screen. If charged negatively, this mesh will repel the electron beam and the screen will remain dark. The input signal leaves a track of positively charged points on the mesh, thus allowing a flood of electrons through, and a trace appears on the screen. This mesh charge remains for some considerable time and the trace is therefore stored.

5.8 The magnetic-tape recorder

The magnetic-tape recorder is beginning to play an increasingly important role in measuring systems, due to the fact that large amounts of data can be very conveniently stored on magnetic tapes.

Figure 5.20 shows the two magnetic heads under which the magnetic tape passes. The tape consists of thin plastics coated with iron-oxide particles which become magnetised if a current flows in the write-head coil. The level of magnetisation depends on the magnitude of the current and, unless magnetic saturation is reached, the relationship is linear.

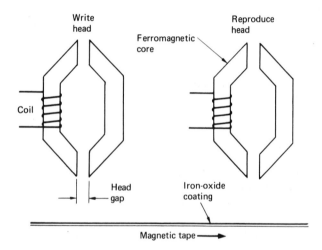

Fig. 5.20 Illustration of the tape-head system of a tape recorder

If information is to be read from the tape, then the reproduce head must be used. Changes in the magnetic flux on the tape induce a voltage in the reproduce-head coil. This voltage is proportional to the rate of change of flux and *not* the magnitude of the flux. The reproduce amplifier must compensate for this differentiation by integrating the signal received from the reproduce head.

There are two main methods of recording on magnetic tape: direct and frequency modulation (FM).

5.8.1 Direct recording

With this method, the intensity of magnetisation is proportional to the amplitude of the signal applied. The larger the applied signal, the more the tape is magnetised, and vice versa. One problem here is that imperfections in the tape give rise to amplitude modulation, and these unwanted signals lead to a poor signal-to-noise ratio (typically 35 dB).

The upper frequency limit is set by the head gap, because, if the wavelength of the signal to be recorded is equal to or less than the gap width, the head will be unable to pick out the variation of magnetisation. The problem could be resolved if the wavelength of the signal were recorded over a greater length of tape, and this means that the tape speed would have to be increased.

5.8.2 FM recording

This method of recording achieves a greater signal-to-noise ratio than the direct method due to the fact that only frequency changes will result in a change of output. Any unwanted amplitude modulation will be ignored, giving a less distorted signal and improved signal-to-noise ratio. A typical carrier frequency is 400 kHz, with deviation of $\pm 40\%$ from this frequency.

5.8.3 Typical magnetic-tape-recorder specifications

Typical magnetic-tape-recorder specifications are given in Table 5.2.

Table 5.2 Typical magnetic-tape-recorder specifications

Mode	Tape speeds	Frequency ranges	Signal-to-noise ratios
FM	1.5 in/s (3.81 cm/s)	0–1 kHz (± 1 dB)	39 dB
	15 in/s (38.1 cm/s)	0–12.5 kHz (± 1 dB)	44 dB
Direct	1.5 in/s (3.81 cm/s)	2.5 Hz–5 kHz (± 3 dB)	35 dB
	15 in/s (38.1 cm/s)	100 Hz–50 kHz (± 3 dB)	39 dB

5.8.4 Digital recording

With the increasing use of microprocessors and home computers, the cassette tape recorder has provided a cheap and readily available means of recording digital data. One system uses two signal frequencies with one combination of these signals corresponding to a logic 1 while another combination corresponds to a logic 0.

A further development of this is to replace the magnetic tape with a large amount of semiconductor memory known as RAM (random-access memory). Some very sophisticated recording systems sample the analogue signal at some high frequency, convert the reading into digital form, and store the results in RAM. When converted back into analogue form, the results can then be played back at a much lower frequency to suit the particular recorder. No doubt as semiconductor memories become smaller with increased capacity, more systems of this type will become available.

5.9 Comparison of recorders

Table 5.3 shows the advantages and disadvantages of each type of recorder and display considered, and also the values of typical frequency responses and response times. It should be remembered that the values shown are typical, and if using a recorder the appropriate operating manual should be consulted.

Exercises on chapter 5

1 What is the difference between a display unit and a recorder? State three examples of each.

2 What is meant by a direct recording instrument? Give an example of an instrument of this type.

3 State three methods of obtaining a permanent trace from recorders which use the basic moving-coil mechanism.

4 A pen recorder produces a record of strain variation in a vibrating aircraft wing section. Determine the input voltage required to produce a trace of 1.5 cm peak-to-peak if the recorder sensitivity is set to 0.2 V/mm. If the strain-measuring-system sensitivity is 1 V per 100 μ-strain, what will be the amplitude of the strain? [3.0 V; 150 μ-strain]

5 Explain the purpose of a u.v. galvanometer matching network.

6 What is meant by optimum damping of a u.v. galvanometer? How does damping affect the step and frequency responses of the u.v. recorder?

7 Select suitable u.v. galvanometers from Table 5.1 and design appropriate matching networks for the following applications:

a) 2 V peak-to-peak 35 Hz output from a pressure-measuring system;
b) 10 V peak-to-peak 180 Hz output from a displacement-measuring system;
c) 1.2 V peak-to-peak 1160 Hz output from a vibration-measuring system.

In all cases, assume a trace size of 15 cm, that the signals are sine waves, and that the signal-source impedance is negligible. [100 Hz; 300 Hz; 2 kHz]

8 A closed-loop recorder is to be used to record cyclically varying temperature changes in a process. If the periodic time is 40 min and the results of 2 cycles are to be included in an A4 sized report, select a suitable chart speed from the following: 5, 10, 20, 50, 100, 200 mm/h. [100 mm/h]

9 State the major advantage of the X–Y plotter over other pen recorders and describe one suitable application.

10 An X–Y plotter is to be used to record a sinusoidal vibration in an aircraft wing section. If the frequency of vibration is 20 Hz, determine the maximum allowable amplitude of trace, using equation 5.1. [1.19 cm]

Table 5.3 Comparison of some recorders and display units

Type	Typical frequency response	Typical response times	Major advantages	Major disadvantages
Mechanical pointers	0 to 0.1 Hz	2 s	Cheap, robust, and suitable for displaying long-term trends	Poor transient and frequency response
Pen recorders	0 to 30 Hz	10 ms	Portable medium-frequency-range device	Low-amplitude traces and limited frequency response
U.V. recorders	0 to 12 kHz (maximum) 0 to 300 Hz (typical)	16 μs 1 ms	Multitrace recording possible	Matching network may be required
X-Y plotters	0 to 2 Hz (3 cm trace) 0 to 1 Hz (6 cm trace)	0.2 s 0.4 s	X-Y recording possible	Very slow response due to inertia in servo-mechanism
CRO	0 to 10 MHz	0.3 ns	High frequency response	Difficulty in getting permanent record
Tape recorder	0 to 12 kHz (FM) 100 Hz to 50 kHz (direct)		Large amounts of data can be stored.	Relatively low signal-to-noise ratios

11 A symmetrical triangular wave is to be recorded on an *X–Y* plotter. If the peak-to-peak value results in a 5 cm trace on the graph paper, calculate the maximum frequency that could be handled by the *X–Y* plotter specified in section 5.6.4.   [25 Hz]

12 A pressure fluctuation in a pipe has a peak-to-peak value of 1.5 bars and frequency 2 Hz and is measured by means of a piezo-electric transducer and a charge amplifier. If the overall sensitivity of the transducer and amplifier is 0.5 V/bar, determine:

a) the size of trace on an *X–Y* plotter if the *Y*-amplifier gain is 0.1 V/cm;
b) the number of cycles recorded on A3 paper (length 38 cm) if the time-base setting is 0.1 s/cm.   [7.5 cm; 7 complete cycles]

13 Explain briefly why the cathode-ray oscilloscope has a good frequency response.

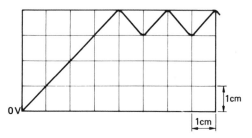

Fig. 5.21

14 Figure 5.21 shows a pressure fluctuation in a compressor displayed on a CRO screen. If the measurement-system sensitivity is 15 mV/bar, the *Y*-amplifier gain setting is 20 mV/cm, and the time-base setting is 10 ms/cm, determine

a) the mean pressure level,
b) the peak-to-peak value of the pressure fluctuation,
c) the frequency of the fluctuation.

Why would the oscilloscope have to be switched to d.c. operation to obtain the trace illustrated?   [4.67 bar; 1.33 bar; 50 Hz]

15 What is the major advantage of the magnetic-tape recorder over other types of recorder?

16 Explain the difference between direct recording and FM recording on magnetic tapes.

17 With reference to the tape-recorder specification in section 5.8.3, calculate (a) the percentage error corresponding to 1 dB error in the frequency response and (b) the 44 dB signal-to-noise ratio as a fraction.   [+12%; −11%; 158:1]

18 If a magnetic-tape recorder is to be used in direct mode, calculate the length of tape used in recording one cycle of the following sinusoidal signals: (a) 25 Hz at a tape speed of 3.8 cm/s and (b) 30 kHz at a tape speed of 38 cm/s.   [1.52 mm; 0.0127 mm]

6   Displacement

6.1 Definition

Displacement is a vector quantity and may be defined as the difference between two positions occupied by a body at different moments in time. Length or distance is a scalar quantity which gives the magnitude of the displacement.

6.2 Units

For a linear motion the displacement will be a length in metres, while for rotational or angular motion it will be an angle in radians.

Until 1960 the metre was defined as the distance between two lines on the standard metre bar kept at Sèvres in France. It was redefined in 1960 as the length equal to 1650763.73 wavelengths in a vacuum of the radiation corresponding to the transition between the levels $2p_{10}$ and $5d_5$ of the krypton-86 atom. The reason for the change was that the new standard could be reproduced in different parts of the world to an accuracy of 1 part in 100 million.

The unit of angular displacement is the radian, which is defined as the angle subtended at the centre of a circle by an arc of length equal to the radius. 1 radian $= 57.3°$.

6.3 Mechanical displacement devices

Dimensional measurement is important in the engineering workshop to determine the size of components being manufactured. Practically, this means the use of devices such as steel rules, slip gauges, micrometer screw gauges, vernier calipers, and dial-test indicators – depending on the size and shape of the component and the accuracy required. Only one example of a mechanical displacement device, namely the dial-test indicator (d.t.i.), will be considered here since most of the items mentioned above will have been covered in some depth in earlier studies on workshop processes.

6.3.1 The dial-test indicator

The d.t.i. illustrated in fig. 6.1 transmits displacement of the test body by means of a plunger which is kept in contact with the body surface by a spring. The signal conditioning of the displacement consists of mechanical amplification by means of gears and a rack and pinion to achieve rotation of a pointer over a scale. A second pointer records the number of rotations of the main pointer.

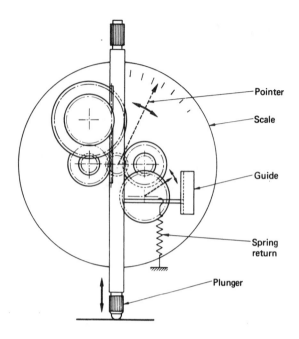

Fig. 6.1 The dial-test indicator

Two points to watch when using the d.t.i. are (i) that the indicator axis lies along the axis of motion, otherwise a component of the displacement and not the actual displacement will be measured, and (ii) that the indicator is initially set up so that it does not reach its limit of travel before the body does. The main pointer scale can be rotated to obtain a zero reading corresponding to some reference displacement, thus enabling displacements from this reference to be measured.

Important uses of the dial-test indicator are in the calibration of electrical displacement transducers, where the indicator is used as an acceptable secondary standard (see section 6.6), and also in force measurement with proving rings (see chapter 9).

Range
Dial-test indicators are available with from 5 mm to 50 mm travel.

Typical specification
Travel   25 mm
Graduation   0.01 mm
Diameter   57 mm
Anvil   ball

Major advantages
a) It is relatively simple to use and read once it is set up.
b) It has an accuracy which is better than most electrical displacement measurement systems – hence its use as a secondary standard.

Major disadvantages
a) It is not suitable for measuring changing displacements, since it is basically a static device.
b) It has no output suitable for recording purposes.

6.4 Electrical displacement devices
Electrical devices are available to measure sizes of manufactured components; for example, the electronic micrometer counts Moiré fringes to produce a digital readout of displacement. However, the emphasis in this section will be on the measurement of displacement resulting from a motion. Typical examples of this type of measurement would be the displacement of a steam-turbine governor valve in order to give an electrical indication of its position, or of the motion of an inlet valve in a diesel-engine test to obtain u.v. recordings of the test results.

6.4.1 Potentiometric displacement measurement
Potentiometric transducers are available for measuring both linear and angular displacements and theoretically give an output voltage whose magnitude is proportional to the displacement. Two important points, mentioned in section 3.7.1, are

i) the load resistance should not be too small compared with the potentiometer resistance, to avoid excessive non-linearity;
ii) the excitation voltage should not exceed the value corresponding to the maximum power rating of the potentiometer.

The potentiometer should preferably be physically connected to the body being displaced, but there are occasions when this is not possible – for example, if the ambient temperature is too high for the transducer. The displacement could then be transmitted through a lever arrangement to a cooler position. Also, if the displacement is too great for the transducer range, it could be reduced by means of a lever as shown in fig. 6.2. Problems of dead-band may arise if there is any play in the linkage, and steps should be taken to minimise this.

Figure 6.3 shows the construction of a linear potentiometer which uses a conductive-plastics track as the resistive element. There is a second low-resistance track which allows electrical contact to be made with the movable wiper, and this provides a variable output voltage dependent on the wiper position. The wiper is mechanically connected to a shaft which moves inside the transducer body, and this shaft in turn is connected to the moving body whose displacement is to be measured.

Both d.c. and a.c. excitation are possible for the potentiometer, and their relative merits have been discussed in chapter 4. D.C. excitation is

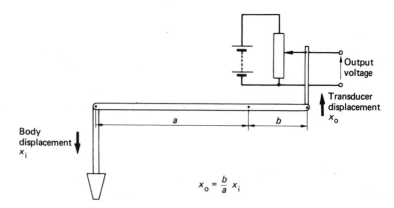

Fig. 6.2 Lever linkage for a potentiometer

$$x_o = \frac{b}{a} x_i$$

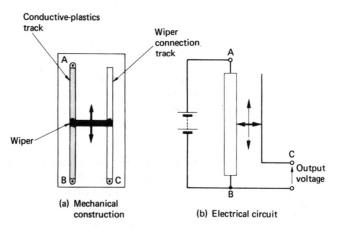

(a) Mechanical construction

(b) Electrical circuit

Fig. 6.3 The linear conductive-plastics potentiometer

the simplest method of signal conditioning, and the magnitude of the d.c. output is usually sufficient to drive recorders or display units directly. A.C. excitation produces a varying amplitude, i.e. an amplitude-modulated signal which requires a demodulator and filter to remove the carrier signal. It is also possible to connect the potentiometer in a bridge circuit which will be balanced when the potentiometer is at its mid position. Examples of both types of signal conditioning are given in section 6.5, where complete measuring systems are considered.

Range
Linear   5 mm to 250 mm
Rotational   up to 340° for a single-turn potentiometer. 3, 5, 10, 15, and 20 turn helical potentiometers are also available.

Resistance range   wire-wound $100\,\Omega$ to $100\,\mathrm{k}\Omega$; conductive-plastics $500\,\Omega$ to $50\,\mathrm{k}\Omega$, depending on the stroke length.

Typical specification (conductive-plastics linear potentiometer)
Electrical stroke length   50 mm
Resistance range $\pm 20\%$ $1\,\mathrm{k}\Omega$ to $10\,\mathrm{k}\Omega$
Standard resistance $\pm 20\%$ $2\,\mathrm{k}\Omega$
Linearity $\pm 0.1\%$ of full scale
Power dissipation at 20°C   2 watts
Operating temperature range -30°C to $+130$°C
Resolution   virtually infinite

Major advantages
a) The high level of output signal could avoid the use of electronic amplifiers.
b) Better linearity than other displacement transducers.

Major disadvantages
a) Friction at the wiper contact will cause wear, although a life of 80×10^6 cycles is quoted for the conductive-plastics potentiometer.
b) Any dirt entering the wiper contact area will affect the output signal. This is particularly applicable to the open construction of some wire-wound potentiometers.

Example 6.1   Using equation 3.3, determine the ratio of load resistance to potentiometer resistance to keep the output-voltage/displacement relationship within a linearity of 1%.

Maximum non-linearity occurs at the mid-point of the potentiometer,

$$\therefore \quad v_{\mathrm{o}} = V\left[2 + \frac{R_{\mathrm{T}}}{R_{\mathrm{L}}}(1 - 0.5)\right]^{-1}$$

$$= V\left[2 + 0.5\,\frac{R_{\mathrm{T}}}{R_{\mathrm{L}}}\right]^{-1}$$

Theoretically, $v_{\mathrm{o}} = 0.5\,\mathrm{V}$ at the mid-point, and to keep within 1% linearity v_{o} should not be less than $0.495\,V$,

$$\therefore \quad 0.495\,V = V\left[2 + 0.5\,\frac{R_{\mathrm{T}}}{R_{\mathrm{L}}}\right]^{-1}$$

$$\therefore \quad 0.495 = \frac{1}{2 + 0.5 R_{\mathrm{T}}/R_{\mathrm{L}}}$$

i.e. $\quad 2 + 0.5\dfrac{R_{\mathrm{T}}}{R_{\mathrm{L}}} = \dfrac{1}{0.495}$

hence $\qquad \dfrac{R_T}{R_L} = 0.0404$

$\therefore \qquad \dfrac{R_L}{R_T} = 24.75$

i.e. the load resistance should be not less than about 25 times the potentiometer resistance for 1% linearity.

Example 6.2 A conductive-plastics potentiometric displacement transducer with the specification given above is to be used to measure the displacement of a process-control valve having a travel of 50 mm. If a 2 kΩ potentiometer is to be used with d.c. excitation, determine (a) the maximum excitation voltage, (b) a suitable value of excitation voltage, (c) a suitable recorder if the maximum frequency of motion is 10 Hz, and (d) the minimum recorder input impedance to keep the transducer linearity to within 1%.

a) From the specification, the maximum power dissipation is 2 W.

$\therefore$ maximum allowable excitation voltage $= \sqrt{PR_T}$

$$= \sqrt{2\,W \times 2000\,\Omega}$$

$$= 63\,V$$

b) Since the maximum excitation voltage is a reasonably high value, the transducer would still give an adequate output voltage if half this value were taken. D.C. supplies of 0 to 30 V are common, and for a convenient transducer sensitivity of 0.5 V/mm select a 25 V d.c. supply.

c) A pen recorder would be suitable.

d) The minimum recorder impedance to keep the linearity to approximately 1% would be 25 times the potentiometer resistance, i.e. $25 \times 2\,k\Omega = 50\,k\Omega$.

6.4.2 L.V.D.T. displacement measurement

The principle of the linear variable-differential transducer (l.v.d.t.), illustrated in fig. 6.4, has been outlined in section 3.10, where it was stated that a core of magnetic material moves inside a coil system to alter the magnetic flux linking the primary and two secondary coils. The two coils are connected in series opposition, to produce an alternating output voltage whose amplitude is proportional to the displacement from the null position. The null position occurs when the core is central, giving equal magnetic linkage between the primary and both secondary windings. When installing l.v.d.t.'s, the core has to be positioned so that the null condition occurs half way between the two limits of displacement.

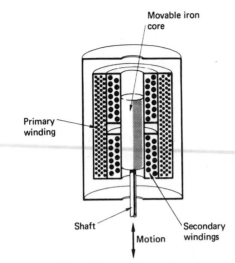

Fig. 6.4 The linear variable-differential transformer

The usual signal conditioning for an amplitude-modulation system is required, namely an a.c. amplifier, a phase-sensitive demodulator, and a filter. Some transducers are available that have all the signal conditioning mounted on the transducer and provide an output voltage in the range 0 to ± 10 V d.c. A ± 15 V d.c. supply is required, and the circuitry includes an oscillator to excite the l.v.d.t. Since the output voltage is typically in the range 0 to ± 10 V, the choice of recorder or display is wide, depending on the frequency of the displacement being within the range of the recorder.

Range
Linear   0.1 mm to 625 mm
Rotational   0 to $\pm 40°$, or to $\pm 60°$ with reduced linearity

Typical specification (linear type)
Nominal linear range ± 25 mm
Input voltage   3 V r.m.s. (nominal)
Frequency range   50 Hz to 10 kHz
Temperature range $-55°C$ to $+150°C$
Transducer sensitivity   16 (mV/mm)/V, i.e. 48 mV/mm for 3 V input
System sensitivity (typical) ± 10 V output from demodulator
Linearity $\pm 0.25\%$ full range
Null voltage   less than 0.5% full-scale output

Major advantages
a) There is no frictional contact between the core and the coils and therefore the l.v.d.t. has a longer life than the potentiometer.
b) Infinite resolution.

Major disadvantages
a) Displacement frequencies of only up to 0.1 of the excitation frequency can be measured.
b) Complex electronic circuitry is required, including an oscillator for frequencies other than mains frequency.

Example 6.3 An l.v.d.t. is to be used to measure the motion of a reciprocating mechanism with a total travel of 40 mm. Using the l.v.d.t. specification above, determine a suitable excitation frequency and recorder if the motion frequencies are (a) 50 Hz and (b) 5 Hz.

The ±25 mm transducer will be suitable for the 40 mm motion.

a) Since the motion frequency is 50 Hz the minimum excitation frequency should be ten times this, i.e. 500 Hz.
   If the a.c. amplifier, demodulator, and filter give an output of the order of ±5 V as indicated in the specification, this is of sufficient magnitude to suit most recorders and the selection would depend on the recorder frequency response. Examination of Table 5.2 shows that the u.v. recorder with a 500 Hz galvanometer would satisfy this criterion, and a suitable matching network would have to be designed.

b) In this case the required excitation frequency would be 50 Hz, which could conveniently be obtained from the mains supply, thus eliminating the need for an oscillator. A step-down transformer would have to be used to obtain the 3 V input voltage.
   The relatively low frequency measurement allows a wider selection of recorders, and a pen recorder would be suitable in this case.

6.4.3 Inductive displacement measurement
Measurement of small displacements by inductive devices will be covered in chapter 11 on vibration measurement, but much larger displacements up to 2500 mm can be measured by transducers of this type.
   The transducer uses a change of coil inductance due to the displacement of an iron core inside the coil. Some transducers have two coils connected in a differential manner and produce a linear variation of inductance with core displacement over a wide range. Thus it is possible to measure greater displacements than with an l.v.d.t., for example.
   There are two possibilities for inductive transducers:

a) to include them in a.c.-excited bridges to produce an amplitude-modulated output voltage;
b) to 'tune' them with a fixed capacitor to obtain a frequency-modulated signal.

Range
0.5 mm to 2500 mm travel

Typical specification
Electrical stroke length 350 mm
Input voltage 210/250 V, 50 Hz
Power consumption 10 W
Output voltage 0 to 10 V maximum into a load of 1 kΩ
Frequency response up to 100 Hz
Linearity ±0.5% of full scale
Resolution infinite
Plunger maximum velocity 5 m/s
Temperature range 0 to 70°C (including electronics)

Major advantages
a) Large displacements can be measured.
b) Friction between plunger and body is insignificant, thus giving a long
   life.

Major disadvantages
a) Not quite as linear as the potentiometer or the l.v.d.t.
b) Frequency range is limited to 0.1 of the excitation frequency of the
   a.c. bridge (not applicable to the FM system).

Example 6.4 The inductive transducer shown in fig. 6.10 (section 6.5.4.)
has the specification quoted above and is operated over only 75% of its
maximum range. If the output voltage is as specified, determine the
trace size if a pen recorder having an input impedance of 20 kΩ and a
sensitivity of 5 mm/V is used.

Would the application illustrated meet the system specification on
frequency response and maximum velocity?

Since only 75% of the maximum travel is achieved, the output voltage
change is 0 to 7.5 V. The pen-recorder input impedance is above the
minimum load specified and therefore the output voltage will be as
stated.

Trace size = 7.5 V × 5 mm/V = 37.5 mm

The upper frequency limit of the measuring system is 100 Hz, and
obviously no hydraulic press operates at frequencies of this order, so the
system's frequency response is quite adequate.

Again, it is difficult to imagine hydraulic presses moving at 5 m/s, but to
analyse this further we have

press travel = 0.75 × 350 mm

= 262.5 mm

= 0.26 m

Assuming a constant speed, to reach the maximum velocity of 5 m/s the
travel would take (0.26 m)/5 m/s = 0.052 s, which is faster than normal
hydraulic presses can achieve.

6.4.4 Capacitive displacement measurement

As mentioned in the previous section, small-displacement measurement will be discussed in chapter 11. There are, however, capacitive displacement transducers available which can measure up to 250mm displacement and give typical outputs of 0 to 10V d.c. Since the signal conditioning is similar to that for the inductive transducers in section 6.4.3, no further details will be given here.

6.4.5 Digital displacement measurement

With the increased availability of digital control systems containing either digital computers or digital logic units, there is a growing need for measurement to be in digital form. Angular and linear displacement transducers are available which give digital information directly, and a typical example is shown in fig. 6.5.

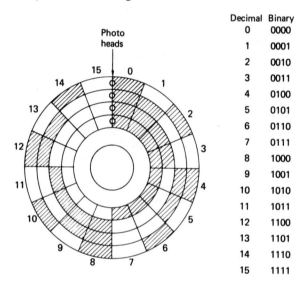

Decimal	Binary
0	0000
1	0001
2	0010
3	0011
4	0100
5	0101
6	0110
7	0111
8	1000
9	1001
10	1010
11	1011
12	1100
13	1101
14	1110
15	1111

Fig. 6.5 The binary-coded encoder disc

The encoder disc has four tracks, corresponding to the four bits in the binary number, and the sixteen segments give a maximum of sixteen digital numbers. The resolution of such a disc is therefore $360°/16 = 22.5°$ and to obtain a better resolution a greater number of tracks on a bigger disc would be required.

Example 6.5 Calculate how many digital numbers can be represented by an eight-track binary disc and determine the resultant resolution.

In general, the maximum number which can be represented is given by 2^n for an n-track disc; so, for eight tracks,

$$\text{maximum number} = 2^8$$
$$= 256$$
$$\text{Resultant resolution} = \frac{360°}{256} = 1.4°$$

Figure 6.6 shows an optical reading system. This has the advantage over mechanical types of being frictionless, with a resulting greater life expectancy. Ambient light can be a problem, and sometimes hoods and lens systems are installed to focus the light beam on to the receiver. Figure 6.6 indicates that if light is blocked on one track then this corresponds to a logic 0 for this particular bit, and if the light is transmitted then a logic 1 results. It is possible to have systems which are the converse of this, but the same principle applies.

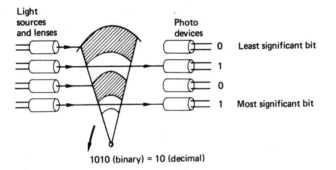

Fig. 6.6 An opto-electronic reading system

One major problem that can arise is that, on changing from one segment to another, one track may change either slightly ahead or behind the other tracks, due to slight misalignment of the receiver heads. This means that for a short period of time the wrong number will be indicated. This is not a serious problem if the angular displacement is only being displayed, but it can cause trouble if the device is being used in a position-control system. The control system will sense an error that does not exist and will try to drive the output to an incorrect position. As an example, in fig. 6.5 consider a change from segment 7 to segment 8, with the most significant bit being slightly ahead of the other three. For a short time all four bits will be displayed, giving a reading of 15 before changing to 8. To reduce these difficulties, a code is used where only one bit at a time changes. The most common example is the Gray code, and Gray-code discs are extensively used in digital position-control systems such as the numerical control of machine tools.

113

Signal conditioning will depend on the application, but Gray code to binary code converters are common, along with other digital logic devices.

The most convenient display for this type of system is the digital display, but results of measurements can be recorded in digital form by the use of punched paper tape or magnetic tape.

Other digital systems
Photoelectric transducers using interference-pattern techniques give a much better resolution than the device mentioned above and, due to their improved accuracy, are found commonly in machine-tool control systems and specifically in the optical micrometer.

6.5 Complete displacement-measuring systems

6.5.1 Measurement of the angular displacement of an aerofoil
Figure 6.7 shows a rotational potentiometer being used to measure the angular displacement of an aerofoil being tested in a wind tunnel. The excitation voltage is of direct form and the output voltage measured between the 'low' and 'wiper' connections is of sufficient magnitude to be fed directly into a pen recorder.

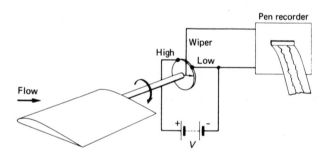

Fig. 6.7 Measurement of the angular displacement of an aerofoil using a potentiometer

6.5.2 Measurement of a hydraulic-jack displacement
Figure 6.8 illustrates a.c. excitation of a linear potentiometer which is connected in an a.c. bridge. The wiper is connected to the jack piston, while the transducer body is attached to the body. With the wiper in the centre of its travel, the bridge is balanced and no output voltage will be obtained. If the wiper is displaced from the null position, the a.c. voltage will be amplified, phase-sensitive demodulated, and filtered to give a d.c. voltage. The measuring system is completed by a u.v. recorder with an associated matching network.

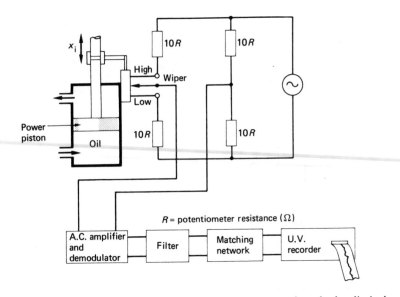

Fig. 6.8 An a.c.-excited potentiometer system measuring hydraulic-jack displacement

6.5.3 Measurement of a steam-turbine valve displacement

The displacement of a steam-turbine governor valve is measured by means of an l.v.d.t. in order to control its position automatically and also to provide a digital display of the valve travel. The complete measuring system is shown in fig. 6.9.

6.5.4 Measurement of a hydraulic-press displacement

Figure 6.10 illustrates the use of a differential inductive transducer connected into an a.c.-excited bridge circuit. Movement of the magnetic core causes an imbalance between the two inductances which results in a bridge output voltage occurring. After amplification and demodulation, the direct voltage is shown being fed into a pen recorder.

6.5.5 A digital angular-displacement system

A Gray-code disc is shown in fig. 6.11 for measuring angular displacement. The digital data is converted to pure binary form before being fed into a logic control system or a digital computer. The display is digital in form and will probably include a binary-to-seven-segment decoder in order to obtain a decimal display.

115

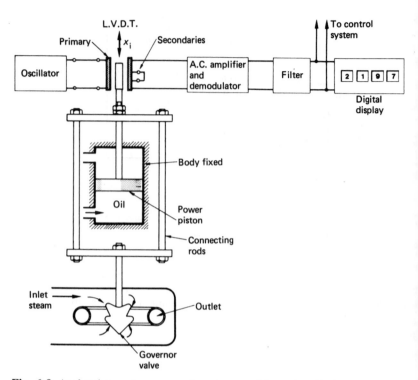

Fig. 6.9 An l.v.d.t. system measuring steam-turbine valve displacement

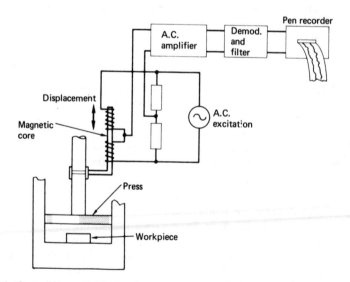

Fig. 6.10 A differential inductive system measuring a press displacement

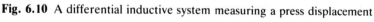

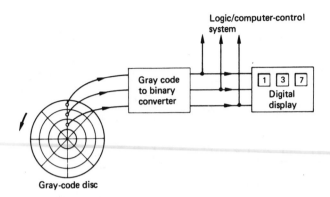

Fig. 6.11 A digital displacement-measuring system

6.6 The displacement transducer in other measurements

Many parameters can be converted to a displacement by the use of appropriate sensing elements; for example, the Bourdon tube and the bellows convert a pressure input into a movement which can be measured by a displacement transducer. There are numerous other examples which will be dealt with in greater detail in the appropriate chapters. To illustrate the point, three examples will be considered here and are shown in figs 6.12 (a), (b), and (c).

Figure 6.12(a) shows the displacement of a Bourdon tube being measured by a potentiometer to give an output voltage proportional to the input pressure.

Level measurement is shown in fig. 6.12(b), where the float converts the level change into a movement of a displacement transducer. One common example of this is the measurement of petrol level in cars.

Finally, fig. 6.12(c) illustrates a seismic-mass transducer used for measuring vibrations, which are transmitted through the spring to the seismic mass. The displacement of the mass relative to the transducer body gives either the amplitude of the vibration or a measure of the corresponding acceleration, depending on the vibration frequency and the values of mass and spring stiffness. Again the displacement is measured electrically.

In all the cases above, the potentiometer has been shown, but obviously other displacement transducers such as the l.v.d.t. could be used.

6.7 Calibration of displacement-measuring systems

All calibration is a comparison with standards, and in practical terms this means a comparison with acceptable secondary standards. The term 'acceptable secondary standards' used in this context is taken to mean acceptable both from the point of view of accuracy and also for convenience or ease of performing a calibration.

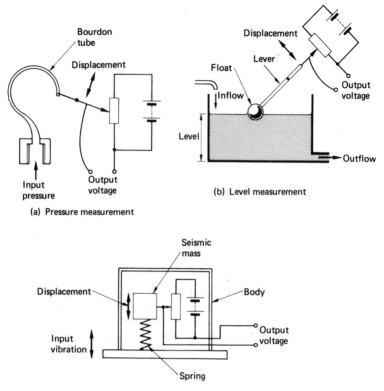

(a) Pressure measurement

(b) Level measurement

(c) Vibration measurement

Fig. 6.12 The displacement transducer used in other measurements

The secondary standards of length are gauge blocks, which are blocks of hardened high-carbon steel with two opposite faces ground and lapped to make them flat and parallel to within certain tolerances. Different grades of gauge block are available, depending on the required tolerance, but a general-purpose block would be typically accurate to within $\pm 1\,\mu m$ at 20°C. Thus, with a set of these secondary standards, a calibration of a micrometer screw gauge for example could be performed. However, these gauge blocks go well beyond the obtainable accuracy of most measuring systems, and so it is more convenient to accept either a dial-test indicator or a micrometer as a practical secondary standard.

Manufacturers of displacement transducers offer calibrators which are basically modified micrometers in order to perform a calibration in the laboratory or at the measurement situation. Figure 6.13 shows a typical calibrator which measures 0 to 25 mm graduated in 0.01 mm increments with a vernier scale readable to 0.002 mm. The anvil is connected either to the core in the case of the l.v.d.t. or to the wiper in the case of the

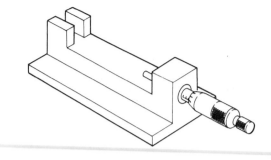

Fig. 6.13 A displacement-transducer calibrator

potentiometer, and a calibration curve of output voltage against input displacement is plotted.

Exercises on chapter 6

1 Compare the advantages and disadvantages of the l.v.d.t. and the linear potentiometer as displacement transducers.

2 A d.c.-excited potentiometer has a sensitivity of 0.8V/mm and is connected directly to a pen recorder which has a sensitivity of 5mm/V. Calculate the overall system sensitivity and the pen movement for a 15mm displacement. [4mm/mm; 60mm]

3 Describe a calibration procedure for (a) a linear potentiometer and (b) a rotary potentiometer.

4 The following results were obtained from a calibration of a l.v.d.t. measuring system using a dial-test indicator for two different gain settings:

Displacement (mm)	−10	−8	−6	−4	−2	0	2	4	6	8	10
D.C. output voltage (V) (gain 1)	−1.18	−0.96	−0.73	−0.49	−0.25	0	0.23	0.47	0.70	0.94	1.17
D.C. output voltage (V) (gain 2)	−2.77	−3.30	−3.95	−2.77	−1.38	0	1.38	2.77	4.26	3.47	2.80

a) Plot the results and evaluate the system sensitivity in both cases. What happened for the gain setting 2?

b) State over what range you would use the transducer with the gain of 2 and evaluate the linearity. [0.118V/mm; 0.77V/mm; 3.75% over ±6mm]

5 Propose an analogue and a digital method of measuring the depth of cut and the traverse motion of the toolpost on a centre lathe.

6 Propose a method of measuring the displacement of a sleeve on a flyball governor system.

7 Suggest a method of measuring the movement of a robotic arm that has both linear and angular motion.

8 How could the depth of insertion of a control rod in a nuclear reactor be measured? [The depth of insertion would be of the order of several metres.]

7   Frequency and angular-velocity measurement

7.1 Definitions

Frequency is the number of cycles, oscillations, or vibrations of a wave motion or oscillation in one second.

Angular velocity is a vector quantity and is the rate of change of angular displacement about an axis. Angular speed is a scalar quantity, which gives the magnitude of the angular velocity.

7.2 Units

The unit of frequency is the hertz, which is equal to one cycle per second. The time taken for one cycle of events is known as the periodic time, T seconds, and the relationship between frequency f and T is

$$f = \frac{1}{T} \text{Hz} \tag{7.1}$$

The unit of angular velocity is the radian per second, although speeds are often quoted in revolutions per second or revolutions per minute (rev/min). Angular velocity ω is given by the equations

$$\omega = \frac{\theta}{t} \tag{7.2}$$

to give an average value, or

$$\omega = \frac{d\theta}{dt} \tag{7.3}$$

to give an instantaneous value, where θ is the angle turned through in radians and t is the corresponding time in seconds.

7.3 Linear and angular motions

There are many cyclic motions which are not rotational, and fig. 7.1 illustrates one such motion which involves a vibrating reed. The term 'frequency' can still be used to describe this motion and gives the number of oscillations about the mean position per second. The device illustrated is actually a simple speed-measuring instrument which is placed on to the body of the rotating machine. Vibrations, which are related to the angular velocity of the rotor, are transmitted through the body and cause the reed to vibrate. When the length of the reed is adjusted to give

121

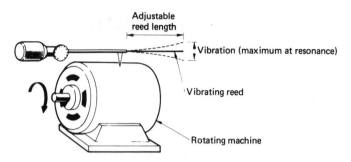

Fig. 7.1 An example of a cyclic linear motion

resonance, the speed of the machine is read off a scale which is calibrated in frequency and corresponding rev/min.

Other examples of linear cyclic motions are the vibrations of a diesel-engine bedplate and the motion of a seismic mass in a vibration transducer.

However, many cyclic motions are rotational, and the measured frequency of the motion will have a linear relationship with the angular velocity. Care should be taken to be certain of this linear relationship, as fig. 7.2 illustrates. The two-pole a.c. generator shown in fig. 7.2(a)

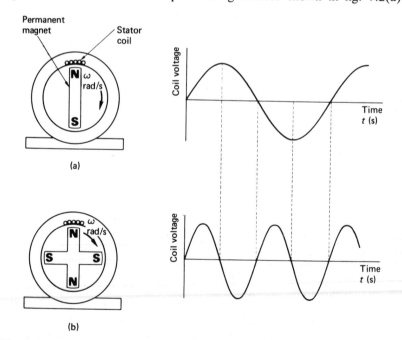

Fig. 7.2 Alternating-voltage generation

produces one cycle of voltage for each revolution of motion, and the equation relating frequency f and angular velocity ω is

$$\omega = 2\pi f \qquad\qquad 7.4$$

Figure 7.2(b), however, shows a four-pole generator which requires only half a revolution to produce one voltage cycle, and in this case equation 7.4 does not apply. The speed of rotation will be half that required by the two-pole machine to produce the same frequency of voltage.

7.4 Measurement of frequency

The measurement technique involved in determining the frequency of a signal will depend on the form of the signal. Digital signals in the form of pulses could require different techniques to analogue signals which may be of sinusoidal form.

7.4.1 The cathode-ray oscilloscope

The CRO, with its calibrated time base, could be used to determine the frequency of both analogue and digital signals. This method involves the determination of the periodic time T, illustrated in fig. 7.3 for both forms of signal, and calculating the frequency from equation 7.1. Care should be taken to ensure that the gain and time-base settings are in the 'calibrate' position before time readings are taken.

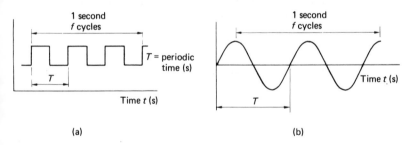

Fig. 7.3 Pulse and sinusoidal waveforms

Example 7.1 A sine wave is displayed on an oscilloscope screen as shown in fig. 7.3 and the distance between the first peak and the fourth peak is found to be 5.7 cm. If the time-base setting is 50 μs/cm, determine the frequency of the sine wave.

Between the first and fourth peak, three cycles have been completed,

$$\therefore \quad \text{distance for one cycle} = \frac{5.7\,\text{cm}}{3}$$

$$= 1.9\,\text{cm}$$

123

$$\therefore \quad \text{periodic time} = 50\,\mu s/cm \times 1.9\,cm$$
$$= 95\,\mu s$$
$$\therefore \quad \text{frequency} = \frac{1}{95 \times 10^{-6}}\,Hz$$
$$= \frac{10^6}{95}\,Hz$$
$$= 10.53\,kHz$$

7.4.2 Lissajou figures

This method of frequency measurement is applicable to sine-wave signals or signals which approach sine waves in form. The signal is compared with a sine wave of known frequency from an oscillator and is displayed on an oscilloscope which is capable of an X–Y display. Other display units with this facility could be used, for example the X–Y plotter if the frequency of the signal is low enough.

Figure 7.4 shows the unknown signal applied to the Y channel and the known variable-frequency signal from an oscillator applied to the X channel of an oscilloscope. The oscillator frequency is varied until a stationary picture known as a Lissajou figure is obtained on the oscilloscope screen. Some adjustment of the X and Y amplifier gains may be necessary to achieve a suitable size of figure.

The shape of the Lissajou figure will depend on the oscillator frequency and also on the phase shift between the two sine waves. Some common

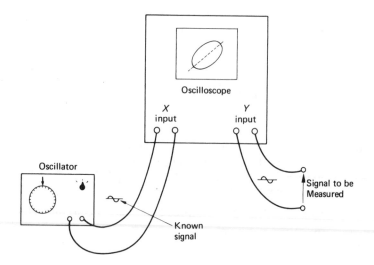

Fig. 7.4 Determination of an unknown frequency by Lissajou figures

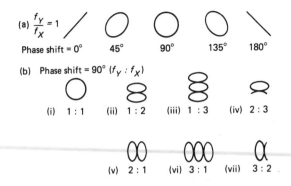

Fig. 7.5 Lissajou figures for different frequency ratios and phase shifts

shapes are shown in fig. 7.5, with the relevant ratio of frequencies $(f_Y:f_X)$. In practice, the signal to be measured will not be a perfect sine wave but may contain harmonics and unwanted noise, so a Lissajou figure such as that illustrated in fig. 7.6 may be obtained. The frequency can still be determined, and the accuracy of measurement will depend solely on the accuracy of the oscillator setting and will be independent of the oscilloscope accuracy.

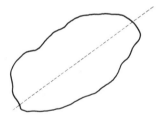

Fig. 7.6 A practical Lissajou figure

7.4.3 Digital methods

Consider the pulse train shown in fig. 7.3(a), where the pulses are recurring at a constant rate. The frequency can be determined by timing the period for one cycle and using equation 7.1, or by counting the number of cycles over a given time. If the frequency of the pulses is f Hz then there will be f cycles in one second. Practically, the time over which the count is taken could be much less than one second for high frequencies or greater than one second for low frequencies. A device for determining the frequency of a digital signal is the timer/counter.

7.4.4 The timer/counter

The timer/counter normally provides a digital display of frequency or time or a count of events over a given time interval. Figure 7.7 shows a typical timer/counter, and the specification quoted below refers to the unit illustrated.

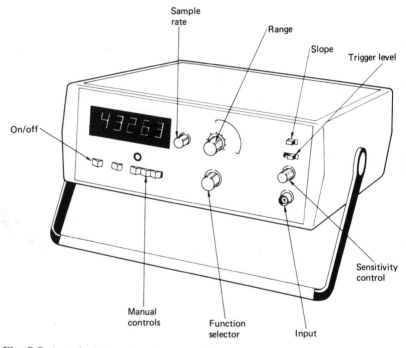

Fig. 7.7 A typical timer/counter

When frequency measurement is selected, the device can accept either pulses or sine waves which are converted to pulses internally. In this mode, the timer opens a 'gate' and counts the number of pulses over a known time interval, at the end of which the 'gate' is closed. The frequency is then evaluated and displayed on the digital display. Gate times vary according to the frequencies being measured.

In the timing or counting mode the device has to be triggered either manually through push-buttons or by an external electrical signal. For example, in the arrangement shown in fig. 7.8 a simple pendulum breaks a light beam and results in the sensor waveform shown. If the timer/counter is arranged to trigger on the negative-going edge (arrowed) then the timer will start on the first edge and stop on the next negative-going edge as illustrated. The periodic time for the pendulum can then be read off the timer display. Note that there will be a small error involved due to the thickness of the pendulum mass.

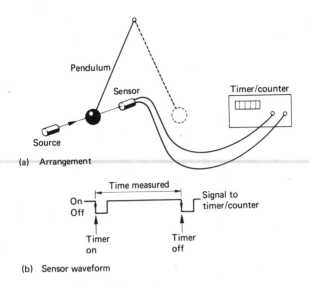

(a) Arrangement

(b) Sensor waveform

Fig. 7.8 Timing a simple-pendulum swing

Due to the relatively high accuracy of the timer/counter, it is sometimes used as a working standard in frequency-dependent systems such as digital speed-measuring systems (see section 7.6).

Range
Frequency range   10 Hz to 35 MHz
Time interval between stop and start signals $1\,\mu s$ to $10^6\,s$
Total number of events (pulses) between stop and start signals   1 to 10^5

Typical specification
Triggering   (a) either positive- or negative-going edge of the input
                        signal
                  (b) manual stop/start
Display   5 digits
Crystal oscillator frequency   1 MHz to within 1 part in 10^6 at a constant
                        25°C; stable to within ± 1 part in 10^5 over
                        a temperature range of 0 to 35°C

Example 7.2 The timer/counter specified above can be operated to measure the frequency of sine waves, and the five-digit display always gives a reading in kHz. For example, 1000 Hz would appear as 1.0000 kHz and 100 Hz would appear as 0.1000 kHz. If the last digit is uncertain, what would be the error expressed as a percentage of the reading for the following frequencies: (a) 500 Hz and (b) 10 Hz?

127

a) A reading of 0.5000 kHz could be 0.4999 or 0.5001 kHz, i.e. ±1 in 5000 or ±0.02% error.

b) A reading of 0.0100 kHz could be 0.0099 or 0.0101 kHz, i.e. ±1 in 100 or ±1% error.

7.4.5 Calibration of frequency-measuring systems

The standard of frequency is the caesium standard, which is the frequency of radiation emitted by a caesium atom when electrons fall between two certain energy levels. In the United Kingdom, this caesium standard is kept at the NPL and is not convenient for general calibration purposes. Use is made of standard-frequency radio transmissions, and the Post Office transmits frequencies of 2.5 MHz, 5 MHz, and 10 MHz from its radio station at Rugby. When the standard frequency is obtained, it is a relatively easy task to obtain submultiples of this by means of frequency-dividing circuits to give a range of known frequencies enabling the measuring system to be calibrated.

Quartz-crystal oscillators are also becoming more common. When electrically excited, these generate signals whose frequencies remain constant to within 1 part in 10^6 or better, and thus go beyond the required accuracy of most frequency-measuring devices.

7.4.6 A complete frequency-measuring system

A frequency-measuring system measuring solely frequency is of little importance to mechanical engineers or technicians, for what concerns them is the parameter whose frequency is being measured. For example, in a speed-measuring system the frequency of pulses may be of prime concern; while in vibration measurement the frequency of the oscillation is important. Details of frequency-measuring systems are therefore given in the sections relating to the appropriate parameter, and in this respect reference may be made to sections 7.8.1 and 11.2.2.

7.5 Analogue methods of angular-velocity measurement

7.5.1 Mechanical tachometers

There are two basic types of mechanical tachometer:

i) the chronometric tachometer, which counts the number of revolutions in a fixed time;
ii) a tachometer incorporating a slipping-clutch mechanism which gives a larger deflection of a pointer as the angular velocity increases.

Both of these types are hand-held on to the rotating shaft and are driven by a rubber cone which is inserted into a suitable dimple in the end of the shaft.

The chronometric type has a pointer which is driven through gearing at a velocity proportional to the shaft velocity for a predetermined time interval, and thus the number of pointer revolutions will be proportional to the shaft velocity. The pointer is automatically held until manually reset, enabling a reading to be taken.

In the case of the slipping-clutch tachometer, the pointer displacement exists only while the mechanism is being driven. A device is therefore available to lock the pointer to allow a reading to be made.

Mechanical loading on the shaft can be a problem, and the errors introduced by a corresponding reduction in speed will depend on the relative sizes of the machine and the tachometer. If the tachometer is small compared to the machine size then the loading effect will be negligible.

Range
Typically   0 to 50000 rev/min

Typical specifications
Three switched ranges   0 to 500 rev/min
                         0 to 5000 rev/min
                         0 to 50000 rev/min

Accuracy   ±1.0%

7.5.2 The a.c. permanent-magnet tachogenerator

Figure 7.2(a) shows an a.c. tachogenerator with a permanent magnet rotating inside a stator coil to induce a voltage, the magnitude of which will depend on the angular velocity of the magnet. There will also be a change in the frequency of the voltage as the speed changes, and this is illustrated in fig. 7.9 where the effect at two velocities – one double the other – is shown. At the higher velocity, both the magnitude and the frequency are double those at the lower velocity.

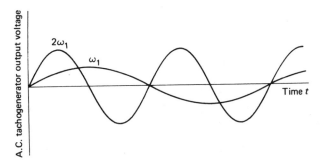

Fig. 7.9 A.C. tachogenerator waveforms for two different speeds

This frequency change affects the impedance or opposition to current in an a.c. circuit and can cause non-linearity in the output-current/input-velocity relationship. A reduction in this effect can be achieved by introducing into the circuit a relatively large series resistance which does

129

not change with frequency. There will, however, be a corresponding reduction in sensitivity.

A rectifying circuit, using a diode bridge and an $R-C$ smoothing circuit as shown in fig. 7.10, can be used to produce a d.c. output voltage suitable to drive directly a pen recorder or a moving-coil meter, for example. The effectiveness in reducing the unwanted ripple is reduced as the frequency decreases.

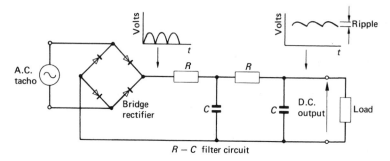

Fig. 7.10 Rectifier and filter circuits used with an a.c. tachogenerator

This type of tachogenerator has the advantage that, since the rotor is a permanent magnet, there are no connections required via carbon brushes and slip-rings. It is therefore more robust and has a greater life expectancy than a comparable d.c. tachogenerator (see section 7.5.5).

Range
Typically 10 to 5000 rev/min

Typical output
60 V r.m.s. at 3000 rev/min, i.e. 1.2 V r.m.s. per rev/s

7.5.3 The a.c. induction generator
The two-phase induction generator is used to obtain an output voltage with constant frequency but whose amplitude changes with speed. Thus the problem of changing impedance with frequency noted with the a.c. permanent magnet tachogenerator is eliminated.

A typical construction consists of two phase windings wound at right angles to each other, with one being excited with alternating current typically at 400 Hz. The alternating field produced by the excited winding induces eddy currents in the moving rotor which is either a squirrel-cage or an aluminium drag-cup rotor. These eddy currents produce a magnetic field in the rotor, at right angles to the main field and linking the second stator winding. Hence a voltage is induced in this winding whose frequency is the same as the excitation frequency and whose magnitude

130

depends on the strength of the rotor field. As the rotor speeds up, the eddy currents increase and induce a voltage whose magnitude is proportional to the shaft speed.

This type of velocity transducer is often found in aircraft control systems, which are energised from a 400 Hz supply.

Range
0 to 4000 rev/min

Typical output
1.0 V at 1000 rev/min

7.5.4 The drag-cup tachometer

The drag-cup tachometer illustrated in fig. 7.11 consists of a permanent magnet which rotates inside an aluminium cup. There is no mechanical linkage between the cup and the magnet, which is attached to the shaft whose speed is to be measured. As the magnet rotates, it induces eddy currents in the aluminium and these create a magnetic field which tries to follow the permanent magnet. The resulting electromagnetic torque is opposed by the torque of a spiral spring, and the cup moves to a position where the two torques balance. The electromagnetic torque due to the eddy currents is proportional to the speed of the magnet; hence the angular position of the cup, indicated by means of a pointer and scale, is proportional to the angular velocity of the shaft.

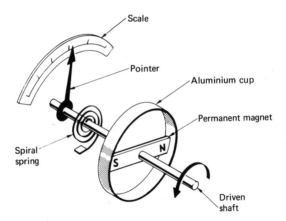

Fig. 7.11 The drag-cup tachometer

A typical application of this type of tachometer is the speedometer in a car, which measures the angular velocity of the car wheels through a geared system.

Range
0 to 10000 rev/min

Typical output
270° pointer movement for 8000 rev/min

7.5.5 The d.c. permanent-magnet tachogenerator
The d.c. tachogenerator is almost physically identical to an a.c. generator with a coil rotating in a permanent-magnet field with one important exception, namely the commutator. Figure 7.12 shows that the conductors of the rotor winding, or armature winding, are connected to insulated copper segments which make contact with spring-loaded carbon brushes. By means of this commutator, the output terminals make contact with conductors which are always in the same position and which produce voltages in the same direction. Thus the output voltage is unidirectional, i.e. a d.c. output, and is proportional to speed.

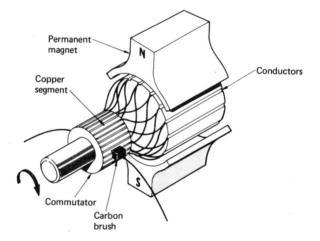

Fig. 7.12 The d.c. permanent-magnet tachogenerator

Range
0 to 8000 rev/min

Typical specification
Output sensitivity   16V per 1000 rev/min
Maximum speed   8000 rev/min
Output linearity up to 5000 rev/min   0.5%
R.M.S. ripple at 3000 rev/min with smoothing time constant of 0.25 ms   0.6%

Summary: analogue instruments

Table 7.1 summarises the characteristics of analogue instruments for angular-velocity measurement.

Table 7.1 Analogue instruments for angular-velocity measurement

Type	Range (rev/min)	Major advantage	Major disadvantage
Mechanical tachometer	0–50000	Direct reading available	Mechanical output not suitable for recording
A.C. permanent-magnet tachogenerator	10–5000	No brush connections	Need for rectification to record signals
A.C. induction generator	0–4000	Constant-frequency output voltage	Need for rectification as above
Drag-cup tachometer	0–10000	Direct visual indication of speed	Mechanical output
D.C. permanent-magnet tachogenerator	0–8000	Direct output voltage available	Need for smoothing circuit to reduce ripple in output

7.6 Digital methods of angular-velocity measurement

All digital methods of measuring angular velocity involve signals in the form of pulses, and it is therefore worthwhile considering in some detail how pulses are generated from the rotating body. One common feature of the methods which follow is that – unlike the analogue devices in section 7.5 – they put almost no mechanical loading on the system.

7.6.1 The electromagnetic pulse technique

This technique requires a toothed wheel made out of magnetic material, for example a normal steel gear wheel, and a coil which is wound round a permanent magnet. Such a system is shown in fig. 7.13, which shows the transducer mounted in close proximity to the toothed wheel, with a gap of 1 mm typically.

As the tip of the tooth arrives opposite the transducer pole-piece, the magnetic circuit is at its best – with a minimum air gap – and maximum flux is produced by the magnet. However, when a root between the teeth is opposite the transducer, the air gap is largest and the magnetic flux is a minimum. These changes in magnetic flux induce an alternating voltage which, for a normal gear wheel, is surprisingly sinusoidal. Since pulses are required for the digital system, the sine wave is fed into a pulse-

133

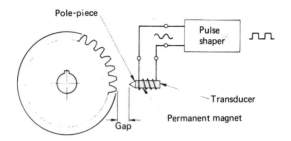

Fig. 7.13 An electromagnetic speed transducer

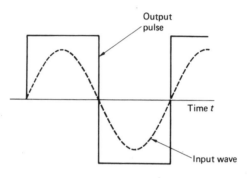

Fig. 7.14 Production of pulses from a sine wave

shaping circuit, for example a Schmitt-trigger circuit, and square waves or pulses are produced as illustrated in fig. 7.14. These can then be fed into a counter to determine the frequency.

Range
Minimum to 120000 rev/min, where the minimum speed depends on the number of teeth used. Typically 1 pulse/s is a minimum frequency.

Typical transducer output
2 volts peak-to-peak at a gap of 1.0 mm (approximately constant amplitude over a wide range of speeds). Larger voltages are possible with smaller gaps.

Example 7.3 A gear wheel with 60 teeth is used in conjunction with an electromagnetic pick-up to measure the angular velocity of a water turbine. If the transducer output is fed into a Schmitt-trigger circuit and then into a timer/counter set to measure frequency, what reading would occur at an angular velocity of 15π rad/s?

$$15\pi\,\text{rad/s} = \frac{15\pi}{2\pi}\,\text{rev/s}$$

$$= 7.5\,\text{rev/s} \ (= 450\,\text{rev/min})$$

$$\therefore \quad \text{frequency of pulses} = 60 \times 7.5\,\text{pulses/s}$$

$$= 450\,\text{pulses/s}$$

$$\therefore \quad \text{timer/counter reading} = 450\,\text{Hz}$$

This is numerically equal to the turbine rev/min, and a 60 tooth wheel is often used to obtain the magnitude of the angular velocity directly in rev/min from a frequency meter.

7.6.2 Opto-electronic techniques

The photocell has been used for a number of years in applications such as counting moving objects on conveyor belts or opening doors when a light beam is broken. Recent developments in opto-electronics have led to the introduction of devices such as the light-dependent resistor (l.d.r.), the phototransistor, and the photodiode, which are light sensors. Light-emitting diodes (l.e.d.'s) have replaced incandescent lamps as light sources, and numerous devices are available with the light source and sensor integrated into one unit. Some l.e.d.'s operate in the visible spectrum, while others operate in the infra-red region.

Figures 7.15(a) and (b) show two typical applications of light sensors being used to measure the angular velocity of a shaft and to count objects on conveyor belts. In fig. 7.15(a), the disc, which is attached to the rotating shaft, has a number of slots in it which cause the light beam to be switched on and off or 'chopped'. The frequency of the resulting pulses

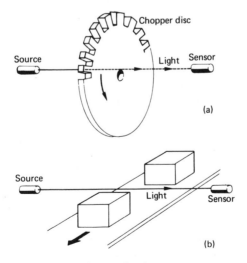

Fig. 7.15 Typical opto-electronics applications

can be determined to give a measure of the angular velocity. In the case of the conveyor belt, fig. 7.15(b), the interruption of the beam triggers a counter to count up each time an object passes.

Two integral units containing both source and sensor are shown in figs 7.16(a) and (b). The slotted opto-switch has a gap of 3mm between the source and the sensor and can be used with a chopper disc to measure angular velocity or with a single tooth to count the number of revolutions. Linear velocity can be measured in a similar manner with the use of a reflective sensor shown in fig. 7.16(b) and a striped pattern of alternate dark and light areas attached to the moving body. The sensor is switched on when the light beam is reflected from a light area and is switched off when no reflection occurs from a dark area. Thus a pulse train is produced whose frequency for equal-length stripes is a measure of the linear velocity.

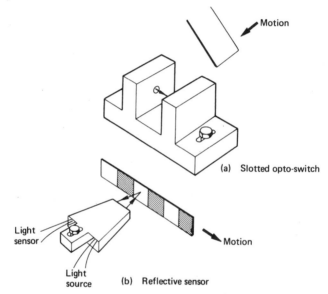

Fig. 7.16 Two integral opto-electronic devices

Striped coding labels on the sides of food packages are now common, and the reflective device would be suitable for reading this coded information. A typical gap for this device is 5mm.

Range
0 to 200000 rev/min

Typical transducer output
4V peak-to-peak from the slotted opto-switch (dependent on supply voltage)

7.6.3 The stroboscope

The stroboscope is a device which consists basically of a gas discharge lamp, normally neon or xenon, which can be controlled to flash at a given rate.

Consider a single distinguishing mark on a rotating shaft which takes a given time to complete one revolution. If the flashing rate can be adjusted so that the time between flashes is equal to the time for one revolution, then the distinguishing mark will be illuminated in the same position at each flash. This will make the mark appear to be stationary, and at this condition the flashing rate will be a measure of the angular velocity,

i.e. $\quad f = f_f$ $\hspace{5cm}$ 7.5

where $\quad f$ = frequency of rotation

and $\quad f_f$ = flash frequency

If the flash rate is adjusted to be slightly higher than the frequency of rotation, then the mark will be illuminated in a position in advance of the previous position and will therefore appear to move slowly backwards. The opposite effect is observed if the flash frequency is slightly lower than the frequency of rotation.

If the flash rate is adjusted so that the time between flashes is half the time for one revolution, then the mark will be illuminated at two positions 180° apart, causing a double image to appear. Conversely, if the time between flashes is double the time for one revolution, then the mark will be illuminated every two revolutions and again a single stationary mark will be observed. This will occur at flash frequencies which are sub-multiples of the rotational frequency; i.e., for a single stationary image,

$\quad f = nf_f$ $\hspace{5cm}$ 7.6

where $\quad n = 1, 2, 3, 4,$ etc.

When using the stroboscope, the technique is to use one single distinguishing mark and to obtain initially multiple stationary images. Then, by reducing the flash rate, the first single image is obtained at the point when the flash frequency equals the rotational frequency. The stroboscope normally has a scale which is marked directly in rev/min and not in Hz.

One other use of the stroboscope is to measure the frequency and magnitude of vibrations in light mechanical systems to which it is not possible to attach electrical transducers. The frequency of vibration can be measured by obtaining a stationary picture, and the amplitude by means of a slowly moving image in front of a graduated scale.

Range
150 to 30000 flashes per minute, although speeds up to 300000 rev/min can be measured indirectly.

Typical specification

Ranges   four overlapping ranges of 1000, 3000, 10000 and 30000 flashes
          per minute
Accuracy   ±1% f.s.d.
Trigger   (a) internal
          (b) external – pulse or sine wave or contact closure

Example 7.4   The speed of a flywheel on an engine is to be measured
using a stroboscope and a single distinguishing mark on the flywheel. At a
setting of 2880 rev/min on the stroboscope, a stationary picture of 3 marks
is observed. What is the speed of the flywheel?

If three marks are viewed then the flash frequency is three times the
rotational frequency,

$$\therefore \quad \text{actual speed} = \frac{2880}{3} \text{rev/min}$$

$$= 960 \text{ rev/min}$$

7.7 Calibration of angular-velocity-measuring systems

Practical calibration of an angular-velocity-measuring system will almost
inevitably involve the use of a timer/counter to measure a frequency
which is directly proportional to the angular velocity. The timer/counter
will have an accuracy far better than the measuring system and so can be
used as a satisfactory working standard.

For digital systems and analogue systems that produce sine waves, the
transducer output could be fed directly into the timer/counter, but for a
system with a d.c. tachogenerator, say, an optical or electromagnetic
pulse system would have to be installed.

7.8 Complete speed-measuring systems

7.8.1 Measurement of the speed of a diesel engine

Figure 7.17 shows a digital system which involves the use of a gear wheel
and an electromagnetic pick-up. The transducer output is amplified with
an a.c. amplifier and then pulse-shaped before being fed into a timer/
counter.

7.8.2 Measurement of the speed of a steam turbine

Figure 7.18 shows an analogue system with a d.c. tachogenerator mounted
on a steam turbine. The smoothing circuit, consisting of a resistor and a
capacitor, is required to remove the ripple on the tacho output due to the
commutator and brushes. Since the output is of relatively large
amplitude, no amplification is required and a direct connection to a pen
recorder is possible.

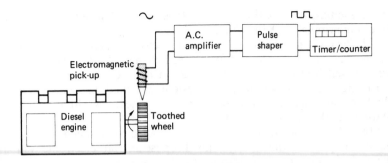

Fig. 7.17 A complete digital speed-measuring system

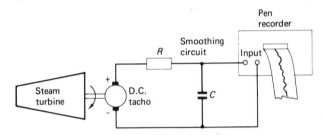

RC = smoothing time constant (s)

Fig. 7.18 A complete analogue speed-measuring system

Exercises on chapter 7

1 Calculate the frequencies of cyclic waves that have periodic times of
(a) 2.5 s, (b) 45 ms, and (c) 20 μs. [0.4 Hz; 22.2 Hz; 50 kHz]

2 Calculate the periodic times of square waves that have frequencies of
(a) 50 Hz, (b) 2.5 kHz, and (c) 30 MHz. [20 ms; 0.4 ms; 0.033 μs]

3 Calculate the frequency of the alternating voltage generated in fig.
7.2(a) when the rotor speed is (a) 3000 rev/min, (b) 1750 rev/min, and (c)
200 rad/s. [50 Hz; 29.2 Hz; 31.8 Hz]

4 An oscilloscope displays a sine wave, and the distance between the
first and third peaks is 4.6 cm. If the time-base setting is 20 ms/cm,
determine the periodic time and frequency of the sine wave. [46 ms;
21.7 Hz]

5 A Lissajou figure as shown in fig. 7.5(b) (iv) was obtained for a known
frequency $f_x = 1.2$ kHz. Determine the unknown frequency. [800 Hz]

6 Compare the advantages and disadvantages of an a.c. and a d.c.
tachogenerator as speed-measuring devices.

7 a) With reference to the specification of a d.c. tachogenerator in
section 7.5.5, calculate the magnitude of the ripple voltage at 3000 rev/
min.

b) Assuming that the ripple varies as the inverse of speed, calculate the percentage ripple at 2000 rev/min.

c) If the tachogenerator is directly connected to a pen recorder, calculate the size of trace obtained for a speed change of 2500 rev/min if the recorder sensitivity is 0.05 cm/V.   [0.29 V r.m.s.; 0.9%; 2 cm]

8 In the electromagnetic pulse system specified in section 7.6.1, calculate the frequency of pulses if the gear wheel has 120 teeth and is rotating at 5000 rev/min.   [10 kHz]

9 State the major advantages of an opto-electronic speed-measuring system compared with other systems. Propose an opto-electronic system for measuring the speed of a turbine shaft.

10 Propose an analogue and a digital system for measuring the speed of an internal-combustion engine. Explain how these systems could be calibrated on the test bed.

8   Strain measurement

8.1 Introduction

The measurement of strain is an increasingly important area of stress analysis. Although there is a wide range of strain-measuring techniques available, as shown in fig. 8.1, some are beyond the scope of this book; hence only electrical strain gauges will be considered.

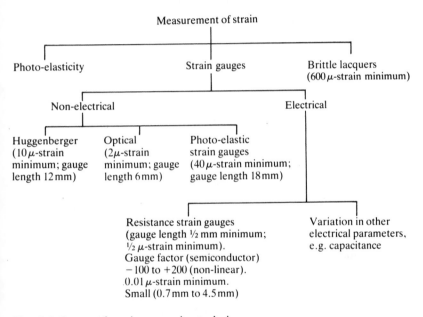

Fig. 8.1 Range of strain-measuring techniques

There are, without doubt, many excellent non-electrical techniques for measuring strain, but the practical advantages of the electrical resistance strain gauges are

a) high accuracy,
b) fast speed of response,
c) good linearity and stability,
d) smallest possible size.

These properties enable them to be applied as transducer elements for the measurement of many physical quantities such as force, pressure, torque, acceleration, and displacement. In addition, their construction and operating principles permit measurement at remote locations, enabling their outputs to be readily applied to control-engineering applications.

8.2 Definition

Strain refers to the relative change in dimensions of a body under stress. It is the ratio of the change in length to the unstressed length of the body,

i.e. $\text{strain } e = \dfrac{\text{change in length}}{\text{unstressed length}} = \dfrac{\Delta l}{l}$

Since strain is a ratio it is dimensionless, but in practice it is often measured in 'microstrain' or 'μ-strain', which means that the ratio is referred to 10^{-6}.

Example 8.1   If a strain is $\dfrac{20 \times 10^{-2}}{100}$, express it in μ-strain.

$$e = \frac{20 \times 10^{-2}}{100} = 2 \times 10^{-3} = 2000\,\mu\text{-strain}$$

8.3 Resistance strain gauges

Resistance strain gauges are transducers which exhibit a change in electrical resistance in response to mechanical strain. They may be of the bonded or unbonded variety, examples of which are shown in figs 8.2 and 8.3.

8.3.1 Bonded gauges

a) Construction

These gauges are bonded, or cemented, directly on to the surface of the body or structure which is being examined; hence any changes in strain in the body are transmitted directly to the gauge material.

Examples of bonded gauges are

 i) fine wire gauges cemented to a paper backing,
 ii) photo-etched grids of conducting foil on an epoxy-resin backing,
iii) a single semiconductor filament mounted on an epoxy-resin backing with copper or nickel leads.

Originally, the only form of gauge construction was the flat grid of continuous wire in one plane, as shown in fig. 8.2(a). However, such a construction results in large dimensions for a high resistance, which led to the development of the helical-coil or wrap-around gauge shown in fig.

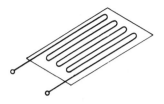

(a) Bonded flat-grid gauge

(b) Bonded helical-wound gauge

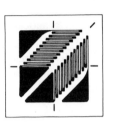

(c) Bonded foil gauges

Fig. 8.2 Bonded strain gauges

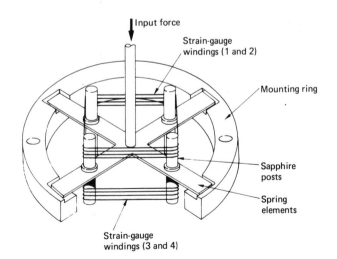

Input force

Strain-gauge windings (1 and 2)

Mounting ring

Sapphire posts

Spring elements

Strain-gauge windings (3 and 4)

Fig. 8.3 An unbonded strain gauge

8.2(b). This uses a continuous wire in two planes and occupies less space for the same resistance.

Some years ago, etching processes led to the development of the foil gauges shown in fig. 8.2(c). These use thin metal foils etched by a photographic process, and as well as small size they have the added advantage that there is no limit to the variety of the forms available.

Semiconductor gauges may be constructed from slices of crystalline materials such as germanium and silicon. These materials, which exhibit a large piezo-resistive effect, are very suitable for strain-gauge construction and are more sensitive than their foil or wire counterparts.

As shown in fig. 8.2, strain gauges can be made up as single elements to measure strain in one direction only, or as a combination of elements, such as rosettes, which permits simultaneous measurements in more than one direction.

b) Classification
The classification of bonded gauges is determined by the resistive and base or backing materials, as well as by the gauge configuration. The following groupings illustrate the classification procedure.

Resistive materials	Base or backing materials	Configurations
Wire	Paper	Single-axial
Foil	Bakelite	Multi-axial
Semiconductor	Polyester	Rosette
	Polymide	Special pattern

A typical strain-gauge classification would thus be 'wire/polyester/single-axial', which would refer to a single-axial wire gauge mounted on a polyester backing.

c) Applications
The choice of gauge for any particular application depends on the type of strain being measured as well as on the environment in which the gauge will be operating.

Static strain, particularly under long-term load conditions, imposes the greatest demands on gauge performance. The selection of the gauge, bonding materials, and connecting wires must be made individually for each application and should provide maximum electrical and dimensional stability, repeatability, and minimum application difficulty.

Dynamic strains, in the absence of static strain, reduce the demands on strain-gauge performance so that materials may be used which enhance the strain sensitivity.

8.3.2 Unbonded strain gauges
A typical unbonded strain-gauge arrangement is illustrated in fig. 8.3, which shows fine resistance wires stretched around supports in such a way

that the deflection of the cantilever spring system changes the tension in the wires and thus alters the resistance of the wire. Such an arrangement may be found in commercially available force, load, or pressure transducers, where the force rod is displaced by the input quantity which in turn deflects the cantilever system.

8.4 Gauge factor

The gauge factor K of a resistance strain gauge is defined as the fractional change in gauge resistance, referred to as 'electrical strain', divided by the strain applied to it,

$$\therefore \quad K = \frac{\text{electrical strain}}{\text{mechanical strain}}$$

$$= \frac{\Delta R/R}{\Delta l/l} \qquad\qquad 8.1$$

Manufacturers supply gauges with a specified gauge factor, which is normally determined by quality-control batch sampling.

Semiconductor gauges have high gauge factors, typically -100 to $+200$, whereas foil and wire gauges have typical values of about 2.

Example 8.2 Determine the uniaxial strain sensed by a $100\,\Omega$ strain gauge if it has a gauge factor of 2 and the resistance change produced by the strain is $1.2\,\Omega$.

Using equation 8.1,

$$\frac{\Delta l}{l} = \frac{1}{K} \times \frac{\Delta R}{R}$$

$$= \frac{1}{2} \times \frac{1.2}{100} = \frac{0.6}{100}$$

$$= 6 \times 10^{3}\,\mu\text{-strain}$$

Example 8.3 A $100\,\Omega$ strain gauge is bonded to a low-carbon-steel bar which is subjected to a tensile load. If the bar has a pre-loaded uniform cross-sectional area of $0.5 \times 10^{-4}\,\text{m}^2$ and Young's modulus for low-carbon steel is $200\,\text{GN/m}^2$, determine the gauge factor if a load of $50\,\text{kN}$ produces a change of $1\,\Omega$ in the gauge resistance.

$$\frac{\text{Stress}}{\text{Strain}} = E$$

$$\therefore \quad \frac{\Delta l}{l} = \frac{\text{stress}}{E}$$

$$= \frac{50000\,\text{N}}{0.5 \times 10^{-4}\,\text{m}^2 \times 200 \times 10^{9}\,\text{N/m}^2} = 0.005$$

Using equation 8.1,

$$K = \frac{\Delta R/R}{\Delta l/l}$$

$$= \frac{1/100}{0.005} = 2$$

8.5 Axes of sensitivity

The active, or principal, axis of the strain gauge is the direction in which the gauge is most sensitive to strain; that is, the direction in which the change of resistance for any given strain is the greatest. The passive, or cross-sensitive, axis is the direction in which the gauge is least sensitive. These axes are illustrated in fig. 8.4.

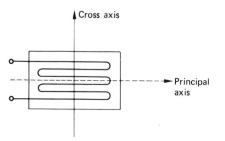

Fig. 8.4 Axes of sensitivity

The cross-sensitivity of the strain gauge may be defined as the ratio of the resistance change ΔR_1 which occurs for a given strain along the cross-sensitive axis to the resistance change ΔR_2 occurring when the same strain occurs along the principal axis.

$$\therefore \quad \text{cross-sensitivity} = \frac{\Delta R_1}{\Delta R_2} \quad \text{for the same applied strain}$$

and is usually expressed as a percentage.

Example 8.4 If the cross-sensitivity of a resistance strain gauge is 1%, and a simple tensile load applied along the active axis produces a resistance change of $0.5\,\Omega$, determine the resistance change if the load acted only along the passive axis.

$$\text{Cross-sensitivity} = \frac{\Delta R_1}{\Delta R_2}$$

$$\therefore \qquad \Delta R_1 = \Delta R_2 \times \text{cross-sensitivity}$$

$$= 0.5\,\Omega \times \frac{1}{100} = 0.005\,\Omega$$

8.6 Bonding techniques

In fixing or bonding a resistance strain gauge to a body with an adhesive, the prime aim is to ensure that the strains in the body are transmitted accurately to the wire or foil of the gauge. In addition there should be good electrical insulation between the gauge and the body, and there should be no 'creep' of the gauge when the body is subjected to strain. Creep is defined as movement of the strain gauge relative to the body to which it is attached.

Potentially, one of the major sources of trouble when using resistance strain gauges is in cementing or bonding the gauges in position. Great care and cleanliness are essential prerequisites to ensure a correctly bonded gauge, and the following procedure is therefore recommended.

a) The surface of the test area should be prepared by first rubbing with a coarse-grade abrasive paper, then finishing with a fine grade in order to produce a definite matt surface.
b) The test surface should then be cleaned with a grease-solvent to remove all traces of grease, oil, or dirt, followed by a final swab with acetone.
c) The underside of the gauge should also be lightly cleaned before installation, using a clean pair of tweezers to handle the gauge.
d) A thin even film of cement should be applied over the area to be gauged and also on the back of the gauge. Over-cementing should be avoided.
e) The strain gauge should then be gently pressed into the cement covering the test area. Care should be taken to avoid the formation of air bubbles.
f) After allowing sufficient time for the cement to set, the insulation or leakage resistance of the gauge to earth should be checked using an insulation tester such as a 'Megger' meter. The value obtained should not be less than $10000\,M\Omega$.

In practice, specific values of insulation resistance are normally quoted by the manufacturer for a given gauge/cement combination.

8.7 Signal conditioning

8.7.1 Introduction

The small changes in resistance of the gauges which occur due to the applied strain may be converted into a voltage by using a Wheatstone-bridge arrangement. The gauge or gauges being used then form part of the bridge circuit along with fixed-value resistors.

8.7.2 Bridge output voltage

It is shown in chapter 4 that the output voltage v_o of an unloaded symmetrical resistance bridge is given by equation 4.15.

i.e. $v_o = \dfrac{V}{4} \times \dfrac{\Delta R}{R}$

where V = the bridge excitation voltage

ΔR = small change in resistance of one element

and R = the resistance of the element before the change

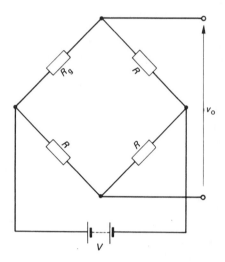

Fig. 8.5 Resistance-bridge circuit used in strain measurement

If the bridge circuit shown in fig. 8.5 is used for the measurement of uniaxial strain, using a single active gauge of resistance R_g, the unloaded output voltage becomes

$$v_o = \dfrac{V}{4} \times \dfrac{\Delta R_g}{R_g}$$

But $\dfrac{\Delta R_g}{R_g} = K\dfrac{\Delta l}{l}$ from equation 8.1

$\therefore$ $v_o = \dfrac{V}{4} \times K\dfrac{\Delta l}{l}$

8.2

or v_o = a constant $\times \dfrac{\Delta l}{l}$

i.e. the output voltage is proportional to mechanical strain for small changes of gauge resistance, the change being usually of the order of 1% of the unstrained resistance value R_g.

It would seem that high sensitivity is possible by using a high gauge factor and/or a large value of excitation voltage. In practice the gauge factor is determined by the type of gauge used, foil and wire gauges having gauge factors of about 2 and semiconductor gauges having typical gauge factors of about -100 to $+200$. The value of the excitation voltage is limited by the gauge manufacturer's recommended operating current or power rating for the gauge, typical values being 15mA and 20mW respectively. The recommendations therefore impose restrictions on the maximum value of excitation voltage that may be used.

Example 8.5 A single $100\,\Omega$ resistance strain gauge, having a gauge factor of 2, is mounted on a steel bar and is connected into a symmetrical bridge circuit. When the steel bar is subjected to a tensile force, the output voltage of the unloaded bridge is 5mV. If the recommended operating current of the gauge is 15mA, determine the value of the mechanical strain.

$$\text{Excitation voltage } V = 2 \times \frac{15\,\text{A}}{1000} \times 100\,\Omega$$

$$= 3\,\text{V}$$

From equation 8.2 we have

$$v_\text{o} = \frac{V}{4} \times K\frac{\Delta l}{l}$$

$$\text{thus} \quad \text{strain} = \frac{\Delta l}{l} = \frac{4v_\text{o}}{VK}$$

$$= \frac{4 \times 5 \times 10^{-3}\,\text{V}}{3\,\text{V} \times 2}$$

$$= 3.333 \times 10^{-3}$$

$$= 3333\,\mu\text{-strain}$$

Example 8.6 If in example 8.5 the gauge factor was 1.9 and the strain input was $4000\,\mu$-strain, determine the value of output voltage corresponding to the new strain input.

Using equation 8.2,

$$v_\text{o} = \frac{V}{4} \times K\frac{\Delta l}{l}$$

$$= \frac{3\,\text{V}}{4} \times 1.9 \times 4000 \times 10^{-6}$$

$$= 5.7\,\text{mV}$$

Example 8.7 A $100\,\Omega$ resistance strain gauge has a recommended operating-power rating of $40\,\text{mW}$. Determine the excitation voltage to be used if the gauge is connected into the symmetrical bridge shown in fig. 8.5.

$$P = \frac{V^2}{R} \text{ watts}$$

$$\therefore \quad V = \sqrt{RP} \text{ volts}$$

$\therefore$ voltage across one gauge is given by

$$V = \sqrt{100\,\Omega \times \frac{40\,\text{W}}{1000}}$$

$$= 2\,\text{V}$$

Since there are two gauges in series in each parallel path, the total excitation voltage will be twice the voltage per gauge, i.e. $4\,\text{V}$.

8.7.3 A.C. or d.c. excitation?

The bridge signal output voltage is rarely large enough for accurate analysis of strain, and the signal is therefore conditioned before being fed to the recording device. The type of conditioner usually depends on the type of measurement to be performed.

There are various techniques for the processing of electrical signals derived from the strain gauges connected in bridge circuits. Three of these methods are as follows:

a) High-frequency a.c. excitation of the bridge. Usually used when only the static component of strain is being measured.

b) D.C. excitation of the bridge. Used when only the dynamic strain is to be measured.

c) When both static and high-frequency dynamic strain are to be measured simultaneously, it appears that the best approach is to employ d.c. excitation of the bridge followed by a low-drift wide-band directly coupled amplifier.

The main advantage of using the a.c.-excited system is that the amplifiers are simple and inexpensive, but the dynamic measurement is restricted since in general the carrier frequencies (usually 1 to 3 kHz) must be approximately ten times the highest frequency component of the dynamic strain.

8.8 Temperature compensation

If a resistance strain gauge is subjected to a changing temperature, its resistance alters, thereby producing an output-voltage variation. This 'apparent strain' due to temperature effects is undesirable and may be overcome either

a) by using self-temperature-compensated gauges,
b) by using 'dummy' gauges, or
c) by arranging the bridge so that temperature effects are eliminated.

8.8.1 Self-temperature-compensated gauges

Self-temperature-compensated gauges may be manufactured from a material such as nickel–copper, nickel–chrome alloy, or treated silicon, whose temperature coefficients of resistance and expansion are such that, when the gauge is bonded to a body of a specified material, the total effect of any temperature change on the gauge resistance is negligible. Unfortunately the temperature compensation is not uniform over the complete operating range of temperature, and some degree of judgement must be exercised in the selection of a particular gauge for minimum response to temperature. This depends not only on the test material but also on the temperature range. For example, a gauge suitable for use on low-carbon steel at high temperatures will, in general, not be self-compensated on the same material at very low temperatures.

8.8.2 'Dummy' gauges

The dummy-gauge method of compensation is accomplished by using two gauges, one in each adjoining arm of the bridge circuit shown in fig. 8.6. Both gauges are matched in terms of gauge factor and resistance and are normally selected from the same production batch.

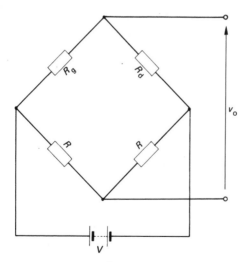

Fig. 8.6 Dummy-gauge compensation

The active gauge R_g is mounted on the test material or body and the dummy gauge R_d is mounted either on the body, in such a way that it does not sense the principal stresses, or on a separate piece of similar material

151

not subjected to the strain being measured but placed near enough so that the 'active' and 'dummy' gauges sense the same temperature. Any temperature variations will have the same effect on both gauges, and the bridge output voltage will remain unchanged as the temperature varies.

8.8.3 Bridge arrangement
This method of compensation is best illustrated by a particular example. Assume that the cantilever shown in fig. 8.7 has two identical gauges

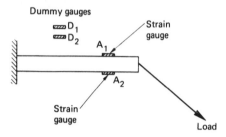

Fig. 8.7 Loaded cantilever with attached strain gauges

mounted on it. If the gauges are connected in adjacent arms of the bridge, temperature compensation takes place since any temperature change will not vary the output voltage. If, however, a load is applied to bend the cantilever, the upper gauge A_1 will be in extension and the lower gauge A_2 will be in compression. The resistance of A_1 will thus rise, while the resistance of A_2 will fall by the same amount. The effect on the output voltage is the same as it would be if A_2 remained unchanged and A_1 changed by twice the amount. The output voltage is thus

$$v_o = \frac{V}{4} \times 2 \times \frac{\Delta R}{R}$$

$$= \frac{V}{2} \times \frac{\Delta R}{R} \qquad\qquad 8.3$$

Remember this for bending strains, dealt with in more detail in section 8.11.1.

8.9 Bridge balancing
In chapter 4 it was shown that the condition for balance of a Wheatstone-bridge circuit is

$$\frac{R_1}{R_2} = \frac{R_4}{R_3}$$

Null-balancing bridges used in strain measurement employ a variable resistor having a scale calibrated in terms of strain. The amount of 'apparent strain' to balance the bridge may then be read directly from the scale. The two most common methods of balancing the bridge are shown in fig. 8.8 and are (i) apex balancing and (ii) shunt balancing.

8.9.1 Apex balancing
A potentiometer, RV1, whose resistance is usually very small compared with the resistance value of the strain guage, is connected at one apex of the bridge circuit shown in fig. 8.8. Movement of the potentiometer wiper increases the resistance of one arm of the bridge while simultaneously decreasing the resistance in the other arm. In this way the bridge may be balanced to compensate for the differences in the resistance values of the gauges and associated fixed resistors.

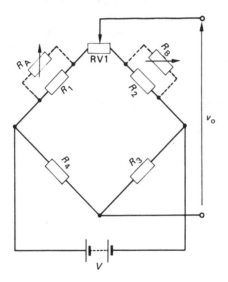

Fig. 8.8 Apex and shunt balancing

8.9.2 Shunt balancing
The values of the shunting or parallel resistors R_A and R_B in the balancing circuit shown in fig. 8.8 are very high compared with the strain-gauge resistance value. This means that the parallel resistors R_A and R_B alter the total resistance of their associated bridge arms by only a small amount.

Example 8.8 Determine the value of the shunting resistor to be placed across a 120Ω strain gauge connected in a symmetrical bridge circuit so that its resistance apparently changes by 1%.

The resistance of the parallel combination of the 120Ω strain gauge and the shunt resistor is

153

$$120\,\Omega - \left(\frac{1}{100} \times 120\,\Omega \right) = 118.8\,\Omega$$

$$\therefore \quad \text{shunting-resistor value} = \frac{120\,\Omega \times 118.8\,\Omega}{120\,\Omega - 118.8\,\Omega}$$

$$= 11880\,\Omega = 11.88\,\text{k}\Omega$$

8.10 Types of circuit arrangement

Bridge circuits can be used in any of the following configurations:

a) One active gauge only. This simple installation is suitable only for dynamic measurements.

b) One active gauge with a temperature-compensating (dummy) gauge. The dummy gauge is mounted so that it is insensitive to principal strains. Dummies mounted on loosely attached plates are used when no transmission of strain is permissible, but temperature compensation may be imperfect. Low output.

c) Two active gauges and two dummy gauges. Moderate output with temperature compensation.

d) Two active gauges for bending strain only. Moderate output, but with temperature compensation, since the active gauges are in adjacent arms of the bridge.

e) Two gauges for longitudinal strain only. High output and temperature compensation.

f) Two active gauges for torsional strain. High output with temperature compensation.

g) Four active gauges. Highest output with temperature compensation.

The bridge circuits are often referred to in the following manner:

i) quarter bridge – where only one strain gauge is used, the other three elements being fixed resistors;

ii) half bridge – where two of the elements are strain gauges, the other two being fixed resistors.

iii) full bridge – where all four elements of the bridge are strain gauges.

8.11 Applications

The following three examples, illustrated in figs 8.9 and 8.10, show how the gauge mounting position and appropriate bridge-circuit connection may be used to measure the following strains in a loaded cantilever:

i) bending strain,
ii) direct strain,
iii) shear and torsional strain.

For each of these three examples, using a half bridge with the notation shown below, it is possible to determine the output voltage in terms of resistance changes when the cantilever is loaded.

In the diagrams shown in figs 8.9 and 8.10 and the following analysis, the notation used is as follows:

A   = active gauge

D   = dummy gauge

e = strain

ΔR = resistance change

R = unstrained resistance of gauge

v_o = output-voltage change

V = excitation voltage

subscripts:

d = direct

b = bending

t = temperature

s = shear

8.11.1 Bending strain
When the cantilever is loaded as shown in fig. 8.7, the gauge A_1 will be in tension and the gauge A_2 will be in compression. The gauges are connected in opposite arms of the bridge, as shown in fig. 8.9(a).

If the temperature increases, both gauges will be subjected to temperature-induced strain. The strains sensed by each gauge will therefore be:

$$\text{strain in } A_1 = e_d + e_b + e_t$$
$$\text{strain in } A_2 = e_d - e_b + e_t$$

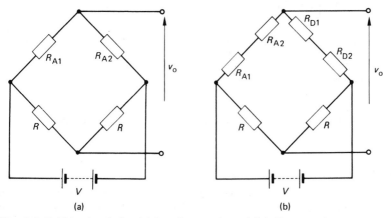

Fig. 8.9 Bridge circuit for (a) bending strain and (b) direct strain

Note the change of sign caused by the difference between compressive and tensile strains.

Using equation 4.16, we find that the output voltage will be given by

$$v_o = \frac{V}{4}\left[\left(\frac{\Delta R_d}{R_{A1}} + \frac{\Delta R_b}{R_{A1}} + \frac{\Delta R_t}{R_{A1}}\right) - \left(\frac{\Delta R_d}{R_{A2}} - \frac{\Delta R_b}{R_{A2}} + \frac{\Delta R_t}{R_{A2}}\right)\right]$$

$$= \frac{V}{2} \times \frac{\Delta R_b}{R} \quad \text{if} \quad R_{A1} = R_{A2} = R$$

i.e. the output voltage is a function of only the bending component in the loaded cantilever, since direct and temperature effects are cancelled.

8.11.2 Direct strain

When the cantilever shown in fig. 8.7 is loaded, the gauge A_1 will be in tension and gauge A_2 will be in compression. The dummy gauges D_1 and D_2 are connected in series in the bridge arm adjacent to that in which the active gauges A_1 and A_2 are connected, as shown in fig. 8.9(b). In this way full temperature compensation is provided as long as the temperatures of the dummy and active gauges remain identical.

The strains sensed by each gauge will therefore be

$$\text{strain in } A_1 = e_d + e_b + e_t$$
$$\text{strain in } A_2 = e_d - e_b + e_t$$
$$\text{strain in } D_1 = \quad\quad + e_t$$
$$\text{strain in } D_2 = \quad\quad + e_t$$

Using equation 4.16 once again, the output voltage will be given by

$$v_o = \frac{V}{4}\left[\left(\frac{\Delta R_d}{2R} + \frac{\Delta R_b}{2R} + \frac{\Delta R_t}{2R}\right) + \left(\frac{\Delta R_d}{2R} - \frac{\Delta R_b}{2R} + \frac{\Delta R_t}{2R}\right)\right.$$
$$\left. - \left(\frac{\Delta R_t}{2R} + \frac{\Delta R_t}{2R}\right)\right]$$

if the unstrained resistances of the active gauges A_1 and A_2 and the dummy gauges D_1 and $D_2 = R$.

$$\therefore \quad v_o = \frac{V}{4} \times \frac{\Delta R_d}{R}$$

which is half the sensitivity of the previous case considered in section 8.11.1.

8.11.3 Shear and torsional strain

The strain gauges are mounted on opposite sides of the cantilever at 90° to each other, as shown in fig. 8.10, and are connected in adjacent arms of the bridge circuit with two gauges acting at 45° to the axis of torsion.

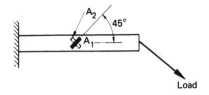

Fig. 8.10 Strain-gauge arrangement for shear and torsional strain

If e'_d is the strain due to direct loading registered at 45° to the beam axis, then, using equation 4.16 again, the output voltage will be given by

$$v_o = \frac{V}{4}\left[\left(\frac{\Delta R_s}{R_{A1}} + \frac{\Delta R'_d}{R_{A1}} + \frac{\Delta R_t}{R_{A1}}\right) - \left(-\frac{\Delta R_s}{R_{A2}} + \frac{\Delta R'_d}{R_{A2}} + \frac{\Delta R_t}{R_{A2}}\right)\right]$$

and, if $R_{A1} = R_{A2} = R$

$$v_o = \frac{V}{2} \times \frac{\Delta R_s}{R}$$

i.e. the same sensitivity as the bending-strain example

8.12 A complete strain-measuring system

The block diagram of fig. 8.11 shows a complete strain-measuring system using a.c. excitation of the bridge. The bridge uses RV1 for resistive apex balancing and a capacitor C which balances out the phase differences caused by the inductance of the connecting wires. The oscillator provides a constant-amplitude high-frequency signal, typically 3kHz, which is used as the bridge excitation voltage and the reference voltage for the demodulation.

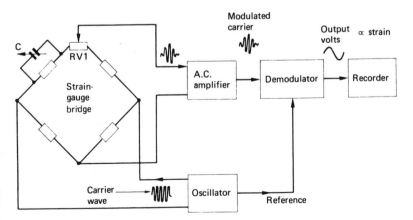

Fig. 8.11 A complete strain-measuring system

When mechanical strain variations are applied, the bridge generates an amplitude-modulated signal which is amplified before demodulation. The demodulator extracts the original form of the physical signal from the amplitude-modulated carrier, and the reference signal ensures that the demodulator is phase-sensitive.

Exercises on chapter 8

1 Define 'gauge factor' of a resistance strain gauge.

If a resistance strain gauge undergoes a 1% change in resistance for an input of 5000μ-strain, determine its gauge factor. If in using the strain gauge you assumed a gauge factor of 1.9, calculate the sense and magnitude of the error in your measurement. [2, +5%]

2 A typical strain-gauge specification is as follows:

a) Gauge resistance $150\,\Omega$ nominal $\pm 0.5\%$
b) Gauge factor 2 with 0.2% tolerance
c) Fatigue life 10^7 cycles
d) Transverse sensitivity 0.3%
e) Drift 10μ-strain/h at 0.7 max. operating temperature

Explain the meaning and significance of each term.

3 A bridge uses one active gauge. The gauge factor is 1.4. If it is attached to a low-carbon-steel bar of diameter 2 cm, what will be the ouput voltage of the bridge when a load of 850 kg is applied in tension? The bridge is supplied at 5 V and is initially balanced. Assume E for low-carbon steel to be $205 \times 10^9 \mathrm{N/m^2}$. [$231\mu$V]

4 A symmetrical bridge uses an active gauge of resistance $200\,\Omega$. Assuming that the bridge has an output impedance of $200\,\Omega$, determine the output voltage indicated on each of the instruments below when the active gauge is subjected to a strain of 350μ-strain. The bridge is supplied at 6 V, and the gauge factor is 2. Instrument (a) has an input impedance of $20000\,\Omega$; (b) has an input impedance of $1000\,\Omega$; (c) has an input impedance of $250\,\Omega$. [1.04 mV; 0.875 mV; 0.583 mV]

5 A particular gauge must not dissipate more than 100 mW during the period of a test. If its resistance is $150\,\Omega$, what is the maximum voltage that can be applied to the bridge? [7.75 V]

9   Force measurement

9.1 Definition

Force is defined as that which changes, or tends to change, the relative motion or shape of a body to which it is applied.

A force can therefore only be recognised by its effects upon the body to which it is applied, and hence these are used to measure the magnitude of the force.

9.2 Units

From Newton's second law of motion we have

$$F = ma$$

where F = force

m = mass of body

and a = acceleration

The unit of force can therefore be defined as the force required to give a unit mass unit acceleration.

In SI units, the unit of mass is the kilogram (kg), the unit of acceleration is the metre per second per second (m/s^2), and the corresponding unit of force is the newton (N).

Force is thus dependent on mass, which is a fundamental quantity, and on acceleration, which is a derived quantity from length and time. The acceleration due to gravity g is a reasonably standard quantity which can be measured accurately, therefore the gravitational force or weight of a known standard mass can be computed to establish a standard weight. A range of standard masses is the basis of a 'dead-weight' tester which is used for the calibration of all force-measuring systems.

9.3 Measurement of force

Force measurement has very wide applications in engineering, ranging from complex in-flight monitoring of forces in aircraft to on-line weighing systems in industry. The latter is a fast-expanding area, since companies who are involved in the pre-packaging of goods for sale by mass must meet government regulations on minimum quantities, while keeping overfill to a minimum to avoid incurring extra cost.

The property of a force (and force-related quantities such as torque and pressure) altering the shape of an elastic component is widely used in different types of force-measuring systems to determine the magnitude of the force. The distortion of the elastic member results in a displacement or strain which can be sensed by means of a secondary transducer. This converts it to an output signal which can be calibrated in terms of applied force. Force-measuring systems therefore incorporate many of the displacement- and strain-measuring transducers discussed in earlier chapters.

Before these types of system are discussed, the oldest types of force-measuring system – the lever and spring types – will be considered.

9.4 Lever-type systems
These determine the unknown force or weight by balancing it against the gravitational force on a known standard mass. Quite often a system of levers is used as an amplifier.

9.4.1 The analytical balance
The analytical balance illustrated in fig. 9.1 is the simplest type of lever arrangement.

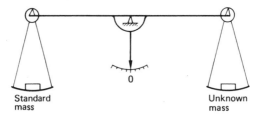

Fig. 9.1 The analytical balance

Null balance is achieved by adding standard masses to the pan to bring the deflection pointer back to zero. For small discrepancies from the standard mass, the scale can be calibrated in fractions of a gram, for direct reading.

9.4.2 Platform scales
A more complex lever mechanism is used in platform scales, one typical arrangement being shown in fig. 9.2.

The system of levers allows the measurement of large forces by means of much smaller standard masses. Coarse adjustment is achieved by adding one of several standard masses; fine adjustment by sliding a mass along a calibrated scale until null balance is achieved. By use of suitable gearing, direct-reading scales can also be produced.

Both of these methods are basically used for 'weighing', i.e. determining the mass of an object. The platform-scale arrangement can be designed for force measurement on dynamometer-type systems.

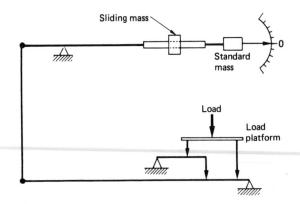

Fig. 9.2 Platform scales

Signal conditioning
Internal signal conditioning consists of levers and pivots for amplification and for converting force moments into angular displacements. No external signal conditioning is necessary.

Range
Static loads from a few milligrams up to a few hundred tonnes on lever-type weighbridges.

Example 9.1 If a set of platform scales has the lever arrangement shown in fig. 9.3, determine the load L on the platform.

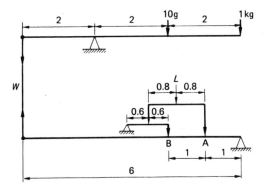

Fig. 9.3 Lever arrangement for example 9.1

With the symmetrical arrangement shown for the platform,

load at A = $L/2$

load at B = $L/4$

161

Therefore, for the bottom lever,

$$6W = \left(\frac{L}{2} \times 1\right) + \left(\frac{L}{4} \times 2\right)$$

$$\therefore \quad W = \frac{L}{6}$$

For the top lever,

$$2W = (0.01\,\text{kg} \times 2) + (1\,\text{kg} \times 4)$$
$$= 4.02\,\text{kg}$$
$$\therefore \quad W = 2.01\,\text{kg}$$
$$\therefore \quad \text{load } L = 6 \times 2.01\,\text{kg} = 12.06\,\text{kg}$$

9.5 Spring-type systems

9.5.1 The spring balance
The spring balance is the simplest form of the spring-type system and has the advantages of being cheap, robust, and easy to use when very accurate measurement is not of prime importance. The load is suspended from a hook attached to a spring which extends an amount proportional to the magnitude of the load. A calibrated scale indicates the load applied.

Signal conditioning
No signal conditioning is necessary for the simple spring balance, but scales with circular calibrated dials require gear-and-lever systems to convert linear motion into angular motion.

Range
Coarse measurement of static loads up to a few kilograms.

Example 9.2 A spring balance incorporates a spring with a stiffness of 10kN/m. If an object hanging on the balance produces a deflection of 40mm, determine the weight of the object and its mass if the local value of g is $9.81\,\text{m/s}^2$.
For the spring,

$$\text{stiffness} = \frac{\text{force}}{\text{deflection}}$$

$$\therefore \quad \text{force} = \text{stiffness} \times \text{deflection}$$
$$= 10 \times 10^3\,\text{N/m} \times 40 \times 10^{-3}\,\text{m}$$
$$= 0.4\,\text{kN}$$

i.e. weight of the object $= 0.4\,\text{kN}$

Also   force = mass × acceleration

$$\therefore \quad mass = \frac{force}{acceleration}$$

$$= \frac{0.4 \times 10^3 \, N}{9.81 \, m/s^2}$$

$$= 40.77 \, kg$$

9.5.2 Hydraulic load cells

The spring used in hydraulic load cells is a metallic bellows as illustrated in fig. 9.4. When load is applied, the compression of the bellows pushes a small amount of hydraulic fluid from the bellows to an indicator via a capillary tube.

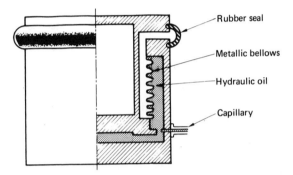

Fig. 9.4 A hydraulic load cell

   These types of load cell are particularly useful for determining the contents of bulk tanks and containers in poor environmental conditions. Their accuracy is unaffected by vessel movement, even when angular or horizontal movement occurs.

Signal conditioning
Bourdon-tube-type indicators are particularly useful and are suitable for remote indication at distances up to 35 m.

Range
Static or low-frequency dynamic loads up to 1000 tonnes.

9.6 Elastic force transducers

In a wide range of force-measuring systems, the primary sensing element is an elastic member. With suitable design, the deflection or strain induced in the member by the external force is proportional to the magnitude of the applied force. Hence, by measuring the displacement or

163

strain with a secondary transducer, the magnitude of the applied force can be determined.

Elastic force transducers are available in many shapes and sizes, the shape and material of construction depending on the application. Figure 9.5 shows some of the more common shapes employed.

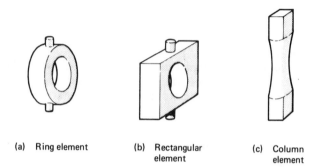

(a)   Ring element   (b)   Rectangular   (c)   Column
                                       element   element

Fig. 9.5 Elastic force transducers

Two of the more common type of device used – the proving ring and the strain-gauge load cell – will be considered.

9.6.1 Proving rings
A load ring, as the name implies, is a short cylindrically shaped element to which the force is applied via integral bosses as shown in fig. 9.6. The

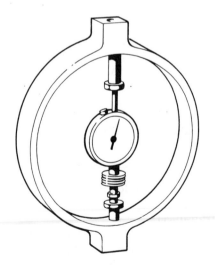

Fig. 9.6 A proving ring

deflection of the ring, usually limited to a maximum of 1% of the outside diameter, is a measure of the force applied.

Load rings have been used for many years, and their characteristics have been refined so that they now have excellent linearity; hence they are used as the primary sensing elements in many force transducers.

High-accuracy load rings are supplied as proving rings suitable for calibrating materials-testing machines, where dead-weight testing is impracticable due to the magnitude of the forces required. The proving ring itself must be subjected to a rigorous calibration check every 2 years by specialist firms licensed by the National Physical Laboratory, according to the procedure laid down in British Standard BS1610:1964. The purpose of the calibration is to check deflection, repeatability, and overload, with an optional check on linearity. The proving ring is then issued with a certificate graded according to the maximum load capacity and the accuracy of the device.

Signal conditioning

Since the output from the load ring, acting as a primary sensing element, is a linear displacement, many of the displacement-measuring techniques described in chapter 6 are used for converting the displacement to an electrical signal. For example, load rings employing l.v.d.t.'s as secondary transducers are available.

For proving rings, the deflection is usually measured using a dial-test indicator (d.t.i.) with a typical resolution of 0.002 mm, rigidly mounted within the ring as shown in fig. 9.6. If a faulty d.t.i. needs to be replaced, it is essential that the proving ring is recalibrated before use.

Range

Proving rings are available in various sizes suitable for measuring forces in ranges from 0 to 2 kN up to 0 to 500 kN. Those employing dial-test indicators are suitable only for static applications, but rings with l.v.d.t. or capacitive displacement transducers are also suitable for dynamic measurement.

Example 9.3 In a calibration test on a load ring, the following results were obtained:

Load (kN)	1.0	1.5	2.0	2.5	3.0	3.5	4.0	4.5	5.0	5.5	6.0
Deflection (mm)	0.065	0.09	0.12	0.15	0.18	0.21	0.24	0.27	0.3	0.33	0.36

Plot a graph of deflection against load and hence determine (a) the sensitivity, (b) the linearity, and (c) the zero error of the device.

From the graph plotted in fig. 9.7 we have

a) Sensitivity is the slope of the characteristic,

i.e. $\quad \text{sensitivity} = \dfrac{0.3 \, \text{mm}}{5.0 \, \text{kN}} = 0.06 \, \text{mm/kN}$

165

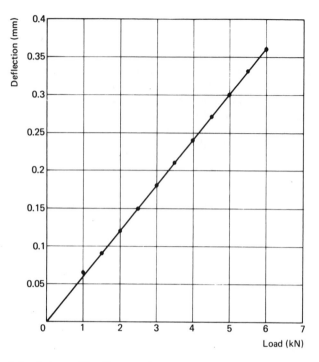

Fig. 9.7 A load-ring calibration graph

b) Linearity is given by the maximum error from the best straight line. This is at 1 kN,

$$\therefore \quad \text{linearity} = \frac{(0.065 - 0.06)}{0.36} \times 100\%$$

$$= \frac{0.005}{0.36} \times 100\%$$

$$= 1.39\% \text{ of full-scale deflection}$$

c) There is no zero error.

9.6.2 Electrical resistance strain-gauge load cells
Load cells have a particularly wide range of applications in the fields of industrial and commercial weighing systems. They are used in such diverse applications as

a) digital display scales used in shops,
b) hot-metal weighing in steel works, and
c) automatic weighing of components on conveyor systems.

In this type of force transducer, the strain induced in an elastic member by the external force is measured. To do this, unbonded gauges and bonded gauges of the foil and semiconductor types described in chapter 8 are all employed by the load-cell manufacturers. Figure 9.8 shows sections through typical load cells using bonded foil gauges.

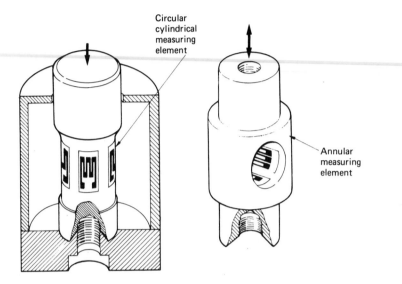

Fig. 9.8 A load-cell cross-section

Gauges are used in the half and full bridge arrangements, with full temperature compensation employed. In addition, if very high accuracy of the order of 0.1% full scale is required, further temperature-sensitive resistors must be incorporated in the circuit as shown in fig. 9.9. These counteract the small temperature-dependent changes which occur in the modulus of elasticity of the material of the primary sensing element.

Elastic members with very high stiffnesses are usually used in load cells, so that the full load strain can be kept to a minimum of the order of 1000μ-strains. This results in generally low sensitivities if solid elastic members are employed in direct tension or compression. The sensitivity can be improved by using elastic members subjected to bending or shear strains when the load is applied. For example, cantilever and load-ring elastic members provide more strain per unit applied load, at the expense of reduced stiffness and frequency range.

General-purpose load cells are supplied for use in tension only, in compression only, and for measuring both tension and compression. Figure 9.10 shows a typical range of load cells and force transducers.

Load cells and force transducers are essentially the same device. The description 'load cell' is usually reserved for transducers which are used

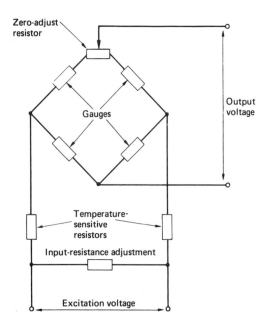

Fig. 9.9 A full-bridge circuit arrangement

for weighing applications in the compression mode, while force transducers are designed for use in a wide range of applications in both tension and compression. They are usually distinguished by their calibration: load or weighing cells are calibrated in units of mass (kilograms) and force transducers in units of force (newtons).

Signal conditioning
Since this type of transducer uses strain gauges in bridge networks, the signal conditioning required is the same as outlined in chapter 8. The manufacturers supply a wide range of portable digital or analogue instruments suitable for direct reading from a wide range of strain-gauge transducers.

Range
Strain-gauge load cells and force transducers are available suitable for nominal loads ranging from 5g or 0.05N up to 1000 tonnes or 10MN. They are suitable for measuring both static and dynamic tension and compression forces to a very high accuracy. The force should always be arranged to be applied parallel to the main force axis, since side loads are a major source of output error, and damage can occur with conventional force transducers if side loading exceeds 10% to 20% of the rated load.

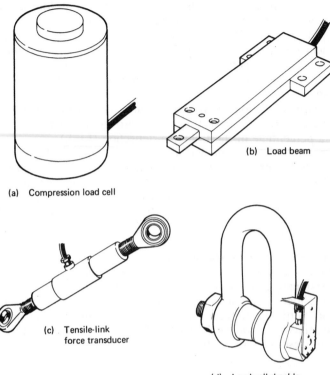

(a) Compression load cell

(b) Load beam

(c) Tensile-link force transducer

(d) Load-cell shackle

Fig. 9.10 Typical load cells and force transducers

Example 9.4 A typical specification for a strain-gauge load cell includes the following items:

a) Nominal full load   3000 kg
b) Overload capacity   4800 kg
c) Bridge configuration   full bridge
d) Supply voltage   maximum 20 V into 240 Ω full bridge
e) Output   nominal 1.80 mV/V at full load
f) Linearity   within ±0.1% full load from best straight line
g) Creep   0.03% in 24 hours
h) Relative static side load   20%

Explain the meaning of each term.

a) Nominal full load is the upper limit of the service range in which the other parts of the specifications apply.
b) Overload capacity is the upper limit of the maximum loading range. If this load is exceeded, the measuring capability of the transducer is permanently damaged.

c) Bridge configuration – full bridge implies that all four arms of the bridge circuit in the load cell are strain gauges.

d) Supply voltage – 20 V is the maximum excitation voltage without the power rating of the strain gauges being exceeded. 240 Ω full-bridge resistance implies that each gauge has 240 Ω resistance.

e) Output – this is the output voltage per unit of voltage excitation at full load.

f) Linearity is the maximum error at any point on the calibration graph measured from the best straight line through the points and expressed as a percentage of the full-scale deflection (f.s.d.)

g) Creep is the maximum change in the rated output voltage at nominal load and reference temperature within a given period of time.

h) Relative static side load is the loading perpendicular to the specified direction of measurement up to which no irreversible mechanical or electrical changes occur. In this case the value will be 0.2 × 3000 = 600 kg.

9.7 Piezo-electric force transducers

Piezo-electric force transducers are particularly useful for measuring high-frequency dynamic forces in test applications such as safety-belt test rigs and shock- and fatigue-testing machines. They are also used for force measurement on automatic stamping, pressing, and welding machines.

The principle of the piezo-electric effect used in this type of transducer has been described in chapter 3. The charge produced on a piezo-electric material, such as quartz, by the application of a force offers an ideal method of measuring the magnitude of the applied force.

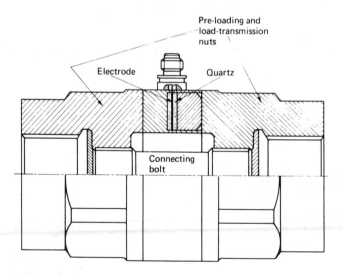

Fig. 9.11 A piezo-electric force transducer

Figure 9.11 shows a cross-section through a typical piezo-electric force transducer. Since a charge is produced only when the quartz discs are compressed, pre-loading nuts as illustrated must be used if the device is also required to measure tensile forces. Tensile forces are therefore sensed as reductions in the initial compressive force applied.

A calibration test is carried out by the manufacturers on all force transducers, and a calibration chart is supplied with each transducer.

Transducers suitable for both compression and tension are calibrated with and without the pre-loading nuts and, provided the stated pre-loading force is applied when remounting the load nuts, the original calibration is still valid.

Signal conditioning

As with all piezo-electric devices, the output is in the form of an electrostatic charge, which must be transformed into a voltage suitable for an indicating or recording instrument. As described in chapter 4, this is achieved by a charge amplifier, which is a d.c. amplifier of very high input impedance and low output impedance.

Range

These transducers are designed to measure dynamic and impact tensile and compressive forces. They have very high resonant frequencies up to 60kHz, which is well beyond the resonant frequency of any mechanical object under test. It may therefore be assumed that any measured variation in force corresponds to the effective variation on the test specimen at any test frequency. Purely static measurements are not possible, due to leakage of the charge; however, with suitable choice of connection lead and charge amplifier, short-duration static tests up to a few minutes are possible.

Force transducers are available with ranges up to 1MN compressive. In the tensile mode with 50% full-range pre-loading on the transducer, tensile forces up to 40% full-scale compressive can be measured – i.e. up to 400kN. The ranges of all the force transducers are summarised in Table 9.1.

Example 9.5 Test equipment for measuring the maximum shock load in a car seat belt incorporates a piezo-electric force link, a charge amplifier, and a recorder. For a certain test carried out at 50km/h, a maximum deflection of 90mm was obtained on the recorder. Determine the maximum value of the shock load if the equipment has the following sensitivities:

Force link   4pC/N
Charge amplifier   10mV/pC
Recorder   1mm/V

$$\text{System sensitivity} = \frac{\text{force-link}}{\text{sensitivity}} \times \frac{\text{charge-amplifier}}{\text{sensitivity}} \times \frac{\text{recorder}}{\text{sensitivity}}$$

$$= 4\,\text{pC/N} \times 10 \times 10^{-3}\,\text{V/pC} \times 1\,\text{mm/V}$$

$$= 40 \times 10^{-3}\,\text{mm/N}$$

$$\therefore \quad \text{maximum force} = \frac{\text{maximum deflection}}{\text{system sensitivity}}$$

$$= \frac{90\,\text{mm}}{40 \times 10^{-3}\,\text{mm/N}} = 2.25\,\text{kN}$$

Table 9.1 Typical characteristics of force transducers quoted in manufacturers' specifications. (Note: the frequency range is limited by load or structure dynamics rather than by transducer dynamics.)

Type	Ranges	Output	Temperature range	Max. transverse load (% of full load)
Balance	0 to 250 g	—	Basically for use at room temperature	Not applicable
Scales	0 to 100 kg	—	,,	,,
Spring balance	0 to 50 kg	—	,,	,,
Hydraulic load cells	Up to 1000 tonne	Depending on pressure transducer applied	,,	50%
Proving rings	Up to 500 kN	Depending on displacement trans. used	,,	Not applicable
Strain-gauge load cells	1000 tonne (10 MN) compression, 500 kN tension and compression	2 mV/V excitation	$-30°C$ to $85°C$	30% low-load range; 10% high-load range*
Piezo-electric	40 kN tension to 1 MN compression	4 pC/N	$-196°C$ to $200°C$	10%

*Shear-force load cells can sustain 100% side loads.

172

9.8 Intelligent weighing systems

'Intelligent weighing system' is the name given to a system incorporating a microprocessor along with the load cell. The microprocessor itself does not provide any improvement in force-measurement accuracy but enables complex control functions to be added. For example, a modern shop weigh-scale provides a digital display of the weight of the goods, the cost per unit weight, and the total cost. The display is retained for a period of time long enough for the shopper to digest the information.

In the manufacturing industries, microprocessors enable load cells on several production lines to be monitored simultaneously. They produce information such as nominal weight, target weights, average weight, standard deviation, and the percentage of samples below the tolerance limit. Additionally, at the touch of a button the current production level of each machine can be called up.

9.9 Complete measuring systems

9.9.1 Measurement of the forces in the cylinder-head bolts of a diesel engine

Figure 9.12 shows an arrangement which enables the forces in the cylinder-head bolts to be measured while the bolts are being tightened. Also, while the engine is running, the additional dynamic forces in the bolts due to the gas load within the cylinder may be recorded.

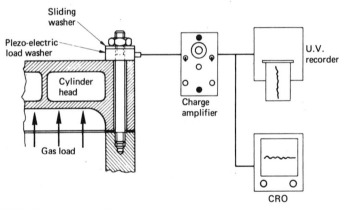

Fig. 9.12 Measurement of bolt stresses

In this type of application, a special-purpose load cell known as a load washer is used to measure the bolt loading. A sliding washer suitably coated with a lubricant must be inserted as shown to avoid damage to the surface of the load washer.

9.9.2 Measurement of the liquid content in a vessel

Figure 9.13 shows a method suitable for the measurement of the mass of a liquid or of a material with a predictable behaviour. The dial indicator can

173

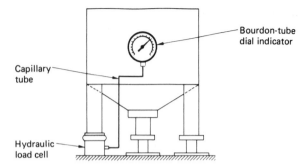

Fig. 9.13 Measurement of hopper content

be suitably calibrated to show weight, volume, or level within the vessel. Remote indication is possible if an electrical pressure transducer is connected to the capillary tube.

For applications where material behaviour is not predictable, a totally supported vessel using three or more load cells could be used. Each load-cell output reading would be passed to a totalising unit which generates a single output for the load-indicating system.

9.9.3 Continuous weighing system

Figure 9.14 shows a method suitable for the recording of the weight of material moving along a conveyor. The signal from the load cell could also be used to control the speed of the conveyor and hence regulate the flow of material to a process. This would then be a closed-loop control system of the type discussed in chapter 13.

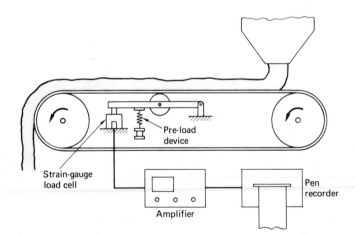

Fig. 9.14 Continuous weighing system

9.10 Calibration of load cells

The calibration technique for virtually all load cells employs a dead-weight tester as illustrated in fig. 9.15.

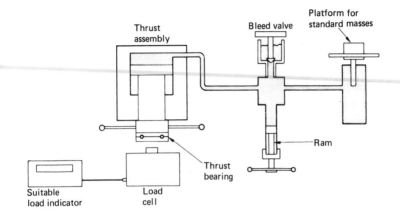

Fig. 9.15 Dead-weight tester

The load cell with its indicator is placed on the loading platform under the thrust assembly, ensuring that the load is applied axially with a minimum of side thrust. All air is bled from the system by opening the bleed valve and screwing in the ram until no more bubbles appear in the reservoir. The ram is then screwed out and the bleed valve is closed. Masses corresponding to the required load are placed on the platform, and pressure is applied by screwing in the ram until the load platform is just floating and can be spun freely. This ensures that the correct load is being applied to the load cell. The starting friction on the load cell can be eliminated by rotation of the thrust bearing, using the handwheel provided. The load-cell reading corresponding to the applied load is noted and the whole procedure is repeated for a range of loads, enabling a calibration graph to be drawn for the transducer.

Ranges of dead-weight testers available
Dead-weight testers incorporating two loading platforms – a low-pressure side and a high-pressure side – are available to cover loading ranges from 25 kg to 25 000 kg (250 N to 250 kN).

Exercises on chapter 9
1 Explain what is meant by an 'intelligent weighing system', giving two typical examples.
2 Describe a procedure for calibrating a load cell.

3 Explain how a piezo-electric force transducer may be used to measure a static load of a few minutes duration.

4 A piezo-electric force link has the following items quoted in its specification:

Measuring range ± 40 kN
Charge sensitivity 20 pC/N
Resonant frequency 37 kHz
Stiffness 21 kN/μm
Overload 10% of full range
Threshold <20 mN

Explain their meaning.

5 A force link, which has a sensitivity of 3.9 pC/N over its full range, is used with a charge amplifier whose sensitivity is set at 10 mV/pC. Determine the forces being measured in three tests in which the following voltages were obtained: (a) -100 mV, (b) 10 V, (c) -75 V. [2.56 N compression; 0.256 kN tension; 1.92 kN compression]

6 One problem in the operation of a jib crane is to monitor its tipping moment, since the rear wheels of the crane will lift if the maximum value of this tipping moment is exceeded. It is therefore necessary effectively to measure the tension in the cable. Propose a method for doing this.

7 It is required to keep the web tension constant in a machine wrapping foil into rolls. Suggest a suitable method for measuring this tension if the transducer can be applied to only one of the rollers over which the foil is passing.

10   Measurement of pressure

10.1 Definition

When any fluid is in contact with a boundary, it produces a force perpendicular to the boundary. The magnitude of the force produced per unit area is called the pressure of the fluid.

Pressure p is therefore defined as the force F per unit of cross-sectional area A,

i.e.   pressure $= \dfrac{\text{force}}{\text{cross-sectional area}}$

$$p = \frac{F}{A}$$

Thus pressure is a derived quantity and has no primary standard; hence other secondary standards must be used for calibration purposes.

10.2 Units

The SI unit of pressure is the pascal (Pa), which is the special name given to the basic unit of pressure, the newton per square metre (N/m^2).

The reference condition known as a 'standard atmosphere' is still used, where

    1 atmosphere $= 101.325\,kPa$

Instead of using the pascal, many manufacturers of pressure transducers are using the bar as the unit on their gauges, where

    $1\,bar = 100\,kPa = 0.987$ atmospheres

One bar can therefore be approximated to one standard atmosphere; hence its adoption by many manufacturers. Other units used by manufacturers, especially for vacuum pressure measurement, are

    1 torr $= 1\,mm$ mercury (Hg) $= 1.333\,22\,mbar$

Table 10.1 gives conversion factors between some of the more common metric pressure units in use.

10.3 Gauge, absolute, and differential pressure

All pressures must be measured relative to some reference. Figure 10.1 shows a general differential-pressure transducer where the sensing element deflects due to the difference between applied pressures p_1 and p_2. If one of the pressures is atmospheric pressure, the pressure sensed is

Table 10.1 Pressure-unit conversion factors

	Pascal (Pa)	Bar	Millibar (mbar)	Torr	Newton/sq. metre (N/m^2)
1 Pa	1	1×10^{-5}	0.01	0.0075	1
1 bar	1×10^5	1	1000	750.062	1×10^5
1 mbar	100	0.001	1	0.750	100
1 torr	133.322	0.001333	1.33322	1	133.322
1 N/m^2	1	1×10^{-5}	0.01	0.0075	1

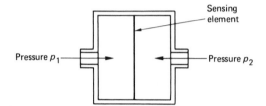

Fig. 10.1 Pressure transducer

known as 'gauge' pressure. This condition is achieved in two ways: either by the reference side being open to atmosphere, i.e. a vented gauge, or by the reference side being sealed at atmospheric pressure, i.e. a sealed gauge. If one pressure is at zero, i.e. the reference side is evacuated and sealed, the pressure sensed is known as 'absolute' pressure.

In each case the units of pressure are not modified but the word 'pressure' should be modified in a statement of the units, as follows: e.g. 'a gauge pressure of 10 kPa', or 'an absolute pressure of 100 bar'.

10.4 Methods of pressure measurement
Pressure can readily be converted to force by letting it act on a known area; therefore many methods of pressure measurement are essentially the same as those described for force measurement in chapter 9. These usually rely on the deflection of some elastic member subjected to the pressure. High-vacuum pressure measurement, however, is not related to force measurement and a variety of special methods are used which will be described briefly later in the chapter.

The first general methods dealt with will be those in which the pressure is balanced against a column of liquid, i.e. manometers.

10.5 Manometers
The action of all manometers depends on the effect of the pressure exerted by a fluid at a depth, and this will be considered in detail before reviewing the different types of manometer.

10.5.1 Pressure at a depth
Three important points about pressure in a fluid are

i) at any point in a fluid, the pressure acts equally in all directions;
ii) fluid pressure always acts normal to any surface; and
iii) the total force acting on a surface, of any shape, in a given direction = (pressure) × (projected area normal to the given direction).

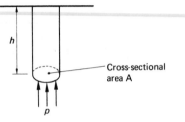

Fig. 10.2 Pressure at a depth

Consider a cylinder of fluid within any quantity of fluid, as shown in fig. 10.2.

Let p = pressure at a depth h

A = cross-sectional area of cylinder

ρ = density of fluid

and g = acceleration due to gravity

The base of the cylinder is in equilibrium under the action of two equal and opposite forces:

force acting upward = fluid pressure × area

$$= pA$$

force acting down = gravitational force on mass of fluid

$$= (\text{mass of fluid}) \times g$$
$$= \rho \times (\text{volume of fluid}) \times g$$
$$= \rho hAg$$

For equilibrium, upward force = downward force

∴ $$pA = \rho hAg$$

$$p = \rho gh \qquad\qquad 10.1$$

Since ρ and g are constant, the pressure at any point in a fluid is directly proportional to its depth in the fluid;

179

i.e. $p \propto h$

Rearranging equation 10.1 we get

$$h = \frac{p}{\rho g}$$

where h is the height of the column of fluid causing the pressure and is known as the *static head*. Typical units for h are millimetres of mercury (mm Hg) or metres of water (m H_2O).

10.5.2 *The U-tube manometer*
The U-tube is the simplest form of manometer and is used for experimental work in laboratories. By suitable choice of liquid, a wide range of pressures can be recorded.

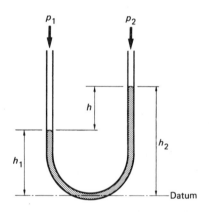

Fig. 10.3 The U-tube manometer

Consider the U-tube manometer shown in fig. 10.3. For equilibrium at the datum point, the total pressure in each limb must be the same. Therefore we have

$$p_1 + \rho_1 g h + \rho g h_1 = p_2 + \rho g h_2$$

where ρ = density of fluid in U-tube

and ρ_1 = density of fluid whose pressure is being measured

Hence $p_1 - p_2 = (h_2 - h_1)\rho g - \rho_1 g h$

But $h_2 - h_1$ = difference in levels, h

∴ $p_1 - p_2 = \rho g h - \rho_1 g h$

or $p_1 - p_2 = (\rho - \rho_1)g h$ 10.2

If the fluid being used in the manometer is mercury and the fluid whose pressure is being measured is air, then the density of mercury is very much greater than ($\gg$) the density of air,

i.e. $\rho \gg \rho_1$

and $p_1 - p_2 = \rho g h$

However if the fluid whose pressure being measured is water, then a comparatively large error can result if the density of the fluid is ignored.

Example 10.1 A U-tube mercury manometer is used to measure the differential pressure in a water-filled pipe, and the difference in the mercury levels in the two legs is 40mm.

a) If the density of water is $1\,Mg/m^3$ and that of mercury is $13.56\,Mg/m^3$, what percentage error is introduced by neglecting the water density in the limb?
b) What is the differential pressure being measured if the local value of g is $9.81\,m/s^2$?

a) From equation 10.2 we have

$$p_1 - p_2 = (\rho - \rho_1)gh$$

Neglecting the density of water, we have

$$p_1 - p_2 = \rho g h$$

$\therefore$ error $= \rho_1 g h$

$\therefore$ % error $= \dfrac{\rho_1 g h}{(\rho - \rho_1)gh} \times 100\%$

$$= \dfrac{1\,Mg/m^3}{(13.56 - 1)\,Mg/m^3} \times 100\%$$

$$= 7.96\%$$

b) Using equation 10.2,

$$p_1 - p_2 = 40 \times 10^{-3}m \times 9.81\,m/s^2 \times (13.56 - 1) \times 10^3\,kg/m^3$$

$$= 40 \times 9.81 \times 12.56\,\dfrac{kg\,m}{s^2\,m^2}$$

i.e.   differential pressure $= 4.93\,kN/m^2$ or $4.93\,kPa$

10.5.3 Well-type manometers
To avoid the inconvenience of having to read two limbs, the well-type manometer shown in fig. 10.4 can be used, which has only one scale arm.

Let A_w = cross-sectional area of well

A_c = cross-sectional area of column

h_w = reduction in level of well

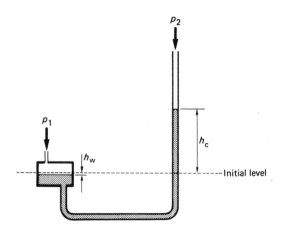

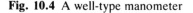

Fig. 10.4 A well-type manometer

h_c = increase in level of column

and p_1 and p_2 = applied pressures

Assuming that the density of the fluid in the well is much less than the density of the mercury, for equilibrium

$$p_1 = p_2 + \rho g(h_c + h_w)$$

i.e. $p_1 - p_2 = \rho g(h_c + h_w)$ 10.3

Now $\dfrac{\text{amount of liquid}}{\text{expelled from well}} = \dfrac{\text{amount of liquid}}{\text{pushed into column}}$

$\therefore$ $$A_w h_w = A_c h_c$$

$$h_w = \frac{A_c}{A_w} h_c$$

Substituting in equation 10.3,

$$p_1 - p_2 = \rho g h_c \left(1 + \frac{A_c}{A_w}\right)$$

If the well area A_w is made large compared with the column area A_c, the zero level moves very little when the pressure is applied and

$$p_1 - p_2 = \rho g h_c$$

Even this small error can also be compensated for by multiplying the scale reading by a factor $A_w/(A_w + A_c)$.

182

The barometer is a well-type single-leg manometer with the column above the mercury evacuated, i.e. at zero absolute pressure, so that the scale reading h_c will be a reading of absolute pressure. High accuracy is achieved by an arrangement which enables the zero level of the well to be set at the zero level of the scale before each reading is taken.

10.5.4 The inclined manometer

The sensitivity of the manometer can be increased by using a lower density fluid such as water instead of mercury as the operating fluid, or by using an inclined manometer as shown in fig. 10.5.

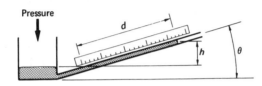

Fig. 10.5 An inclined manometer

Let θ = angle of inclination of manometer,

h = vertical increase in head

and d = movement of column along limb

Then $d = \dfrac{h}{\sin \theta}$ 10.4

Example 10.2 If an inclined manometer as illustrated in fig. 10.5 has an angle of inclination of 10°, determine the increase in sensitivity.

Using equation 10.4,

$$d = \frac{h}{\sin 10°} = 5.76h$$

∴ increase in sensitivity = 5.76

Signal conditioning

In manometers, signal conditioning is achieved by the pressure being converted into a movement of liquid in a column. The column is calibrated and generally read by the operator. Automatic tracking of the level can be achieved by means of a float and a differential transformer, to produce an electrical signal.

Range

Manometers are used for static pressure measurement as follows:
U-tube and well – 1mm mercury to 1m of water or mercury, i.e. approximately 1mbar to 100mbar or 1.5bar.

Inclined – 1 mm to 300 mm of water, i.e. approximately 0.1 mbar to 30 mbar.

10.6 Elastic pressure transducers

Many pressure transducers use elastic primary sensing elements as described in chapter 3, the most common being the Bourdon tube, the bellows, and the diaphragm.

10.6.1 The Bourdon-tube pressure gauge

A Bourdon tube is a long thin-walled cylinder of non-circular cross-section sealed at one end, made from materials such as phosphor bronze, steel, and beryllium–copper. A pressure applied to the inside of the tube causes a deflection of the free end, proportional to the applied pressure.

The most common shape employed is the C-type, as illustrated in fig. 10.6. Increased sensitivity can be achieved by using spiral and helical-shaped tubes.

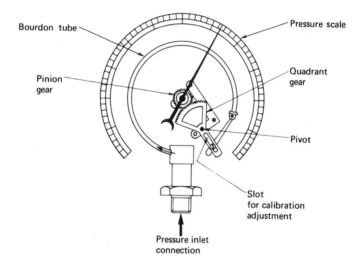

Fig. 10.6 A Bourdon-tube pressure gauge

Signal conditioning

In the simplest form of mechanical pressure gauge, illustrated in fig. 10.6, the displacement is converted into a pointer rotation over a scale by means of a gear-and-lever system. Remote indication of pressure can easily be achieved by using any of the displacement transducers described in chapter 6, e.g. potentiometers and linear variable-differential transformers (l.v.d.t.'s).

Range
Static and low-frequency pressures up to 500MN/m². The frequency range is limited by the inertia of the Bourdon tube if electrical displacement transducers are used.

10.6.2 Diaphragm pressure transducers
Flat diaphragms are very widely employed as primary sensing elements in pressure transducers using either (a) the centre deflection of the diaphragm or (b) the strain induced in the diaphragm. They can be conveniently fabricated as flush-mounted sensing elements, providing a clean smooth face, ideal for use in dirty environments and for surface-pressure sensing.

For high-pressure transducers, very stiff diaphragms must be used to limit the centre deflection to less than one third of the diaphragm thickness, otherwise non-linearities result. For lower pressure ranges, up to a few bars, beryllium–copper corrugated diaphragms and bellows are also used to give the higher sensitivity required.

a) Displacement sensing
Potentiometric, inductive, and capacitive displacement transducers as described in chapter 6 are all used to sense the diaphragm displacement. A schematic diagram of a typical inductive type of pressure transducer is shown in fig. 10.7.

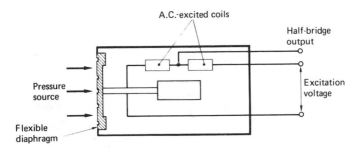

Fig. 10.7 An inductive pressure transducer

Capacitance pressure transducers are often used in microphones for noise measurement and are described in chapter 11.

Signal conditioning
This depends on the type of transducer used.

Range
Typically for l.v.d.t. displacement measuring, the pressure range is up to 200bar. Dynamic pressure measurement can also be carried out at fre-

quencies up to 0.5kHz for low-pressure transducers of 0.1bar range, or a maximum of 15kHz for a 200bar transducer.

b) Strain sensing

The diaphragm acts as a force-summing device, converting the applied pressure into a force which causes strain in the material of the diaphragm. Four main strain-measuring configurations are used:

i) Unbonded strain gauges as illustrated in fig. 3.6, where the strain is applied directly to the fine wire.

ii) Bonded strain gauges, using foil-type gauges bonded to the diaphragm. This type is unsuitable for elevated temperatures above about 85°C, since the adhesion of the gauge to the surface of the diaphragm may vary, causing slight repositioning and hence inaccuracies.

iii) Sputtered or vapour-deposited gauges overcome the variable-adhesion problem at elevated temperatures. In this type of transducer, the gauges with their insulation are vapour-deposited on to the diaphragm, ensuring a permanent bond. The sputtered-gauge pressure transducer is very stable with time over a broad temperature range from −40°C to 120°C.

iv) Semiconductor strain gauges use a single-crystal silicon diaphragm, with the strain gauges diffused directly into the crystal as illustrated in

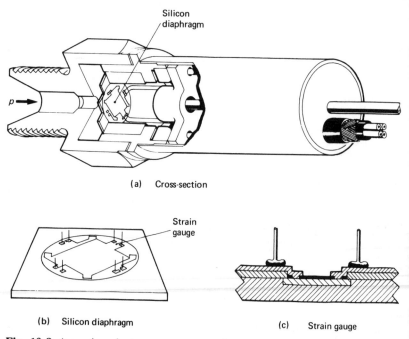

(a) Cross-section

(b) Silicon diaphragm

(c) Strain gauge

Fig. 10.8 A semiconductor pressure transducer

186

fig. 10.8. Stress levels within these devices are lower and sensitivities are higher than in other types of strain-gauge transducers. They are very rugged transducers with a high resistance to shock and vibration and, by using special p.t.f.e. cable, can be used in the temperature range $-40°C$ to $+150°C$.

Note: All of the temperature ranges can be extended by using special air- or water-cooled transducers.

Signal conditioning

All of the strain-sensing transducers employ Wheatstone-bridge networks, and generally full-bridge arrangements are used to allow automatic temperature compensation with high sensitivity.

Transducers are now available incorporating integrated signal-conditioning circuits within the same housing as the sensing element. These devices include internal temperature compensation, voltage regulation, and full signal conditioning by an operational amplifier giving a low-impedance 10V maximum output.

Range

Strain-gauge pressure transducers have a wide operating range from 50mbar using the high sensitivity of miniature semiconductor transducers up to 2000bar using vapour-deposited-gauge transducers. They are suitable for measuring static and dynamic pressures. The individual ranges of the transducers are summarised in Table 10.2 at the end of the chapter.

Note   An important requirement for all pressure transducers suitable for measuring dynamic pressures is that they should have a low sensitivity to acceleration effects – i.e. low '*g*' sensitivity – since high accelerations can produce effects in the primary sensing element similar to pressure variations. This is usually achieved by using low-mass high-stiffness primary sensing elements.

Example 10.3   A pressure transducer has a sensitivity of 2.5mV/bar when correctly energised. The output impedance of the transducer is 250Ω and it is connected to a galvanometer with an internal resistance of 50Ω. If the galvanometer has a sensitivity of $10mm/\mu A$ and the galvanometer spot deflects 75mm, determine the magnitude of the pressure being measured.

The circuit can be represented as shown in fig. 10.9.

$$\text{Current } I \text{ required for 75mm deflection} = \frac{75mm}{10mm/\mu A}$$

$$= 7.5\mu A$$

$$\text{Total circuit resistance } R = 250\Omega + 50\Omega$$

$$= 300\Omega$$

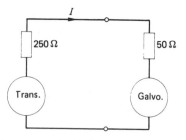

Fig. 10.9 The circuit for example 10.3

Voltage V to give current $I = IR$

$$= 7.5 \times 10^{-6}\,\text{A} \times 300\,\Omega$$

$$= 2.25\,\text{mV}$$

$\therefore$ pressure being measured $= \dfrac{\text{output voltage}}{\text{transducer sensitivity}}$

$$= \frac{2.25\,\text{mV}}{2.5\,\text{mV/bar}}$$

$$= 0.9\,\text{bar}$$

10.7 Piezo-electric pressure transducers

Pressure transducers using the piezo-electric effect use a similar design to the quartz load cells described in chapter 9, the quartz discs being compressed by a diaphragm which is in direct contact with the pressure being measured. The high sensitivity of the quartz-crystal modules permits the transducers to be manufactured in extremely small sizes, e.g. cylinder pressures in petrol engines can be measured using a spark plug modified to include the transducer.

One outstanding feature of quartz transducers is their high sensitivity; e.g. a transducer designed for 250 bar maximum pressure will give 7.5 pC for a pressure variation of 0.1 bar, which with suitable signal conditioning can produce a 750 mV output signal.

Example 10.4   The waveform obtained with a piezo-electric pressure-transducing system is shown in fig. 10.10. If the transducer sensitivity is 60 pC/bar and the charge-amplifier sensitivity is set at 20 mV/pC, determine (a) the mean pressure and (b) the peak amplitude of pressure fluctuations.

$\begin{aligned}\text{Overall} \\ \text{system sensitivity}\end{aligned} = \begin{aligned}\text{amplifier} \\ \text{sensitivity}\end{aligned} \times \begin{aligned}\text{transducer} \\ \text{sensitivity}\end{aligned}$

$$= 20\,\text{mV/pC} \times 60\,\text{pC/bar}$$

$$= 1200\,\text{mV/bar}$$

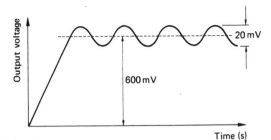

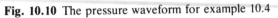

Fig. 10.10 The pressure waveform for example 10.4

a) Mean pressure $= \dfrac{600\,mV}{\text{system sensitivity}}$

$ = \dfrac{600\,mV}{1200\,mV/bar} = 0.5\,bar$

b) Peak amplitude of pressure fluctuations $= \dfrac{10\,mV}{\text{system sensitivity}}$

$ = \dfrac{10\,mV}{1200\,mV/bar}$

$ = 8.33\,mbar$

Piezo-electric pressure transducers can be operated at temperatures up to 240°C, although care must be taken to compensate for zero-drift with temperature. Special water-cooled transducers are available, as illus-

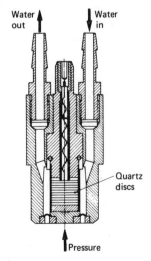

Fig. 10.11 A water-cooled piezo-electric pressure transducer

189

trated in fig. 10.11, and these are particularly useful for high-temperature applications.

Signal conditioning
As for all piezeo-electric devices, a special high-input-impedance charge amplifier is required.

Range
Short-duration static pressures can be measured, but piezo-electric transducers are more generally used for dynamic-pressure applications such as testing engine compression, blast-pressure measurement, and testing ammunition where the very high frequency range up to 150 kHz is suitable. Pressures up to 5000 bar can be measured.

10.8 Measurement of vacuum
Vacuum pressures are those which are below atmospheric. With modern vacuum pressure systems it is possible to obtain pressures from 1000 mbar (approximately 1 atmosphere) down to 10^{-12} mbar; however, no single pressure transducer is available which covers this full range, and a suitable pressure transducer must be chosen to cover the range required for a particular application. Down to 1 mbar it is possible to use some of the techniques so far discussed, e.g. manometers and diaphragm-type transducers; below this level, however, very different techniques must be used.

10.8.1 Capsule gauges
Capsule gauges employ a variation of the diaphragm sensing element. They use two corrugated diaphragms joined at the edges or more often manufactured by vacuum forming into a capsule as shown in fig. 10.12. Vacuum gauges use a sealed evacuated capsule within the measuring chamber. Pressure variations cause the capsule dimensions to change.

Signal conditioning
In the mechanical form, the dimension changes in the capsule are transmitted via mechanical linkages to a rotary pointer.

In electrical devices, a potentiometer is used to convert displacement to an electrical signal.

Range
The pressure range is from atmospheric down to 0.5 mbar. The mechanical gauges are suitable for static measurement only; the electrical devices can be used for low-frequency pressure variations.

Fig. 10.12 A capsule diaphragm

10.8.2 The McLeod gauge

The McLeod gauge employs a special type of manometer arrangement, fig. 10.13. The 'Vacustat' series of gauges manufactured by Edwards High Vacuum Ltd are designed for industrial as well as laboratory use.

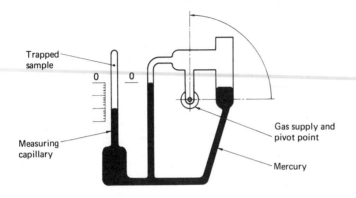

Fig. 10.13 The McLeod gauge

The gauge is positioned with the measuring capillary horizontal and connected to the gas pressure to be measured. To record the pressure, the measurement head is rotated through 90° into the position shown in fig. 10.13. The rotation causes the mercury to move within the glass tubes and isolate a fixed volume of the gas whose pressure is to be measured. The trapped gas is then compressed into the closed measuring capillary and, when the mercury in the open capillary is level with the zero mark on the measuring scale, the gas pressure is given by the height of the mercury column on the calibrated scale.

This type of gauge is unsuitable for measuring the pressure of a gas containing condensable vapours, as they tend to condense out during the compression process.

Signal conditioning
Signal conditioning is carried out by the capillary tubes which convert the gas pressure into a mercury height.

Range
Suitable for static vacuum pressure measurement from 10 mbar down to 10^{-3} mbar.

10.8.3 Thermal-conductivity gauges
The thermal conductivity of a gas is pressure-dependent, and thermal-conductivity gauges operate by measuring the change in this conductivity as the gas pressure varies. This can be achieved in two ways: by Pirani gauges and by thermocouple gauges.

a) Pirani gauges

These gauges operate by measuring the changing resistance of an electrically heated filament as its temperature changes due to the cooling effect of the surrounding gas. The transducer consists of a gauge head of metal or glass incorporating a hairpin filament of fine tungsten wire heated to an operating temperature of 160°C. The gauge head is connected to a measuring instrument which supplies the heating current for the coil, the signal conditioning, the recording scale, and often output terminals for connection to an external continuous recorder. A typical arrangement is shown in fig. 10.14.

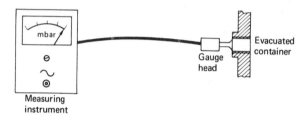

Fig. 10.14 The Pirani-gauge arrangement

Signal conditioning
The transducer produces a small change in resistance proportional to the gas pressure, which is sensed by means of a Wheatstone-bridge circuit.

Range
The gauges are suitable for measuring static vacuum pressures from 100 mbar down to 10^{-4} mbar, with response times in the order of 50 ms.

b) Thermocouple gauges

These gauges use a thermocouple attached closely to a heated filament to measure its change in temperature due to the surrounding gas-pressure variations. This type of gauge also consists of a gauge head and a measuring instrument. The gauge head used is usually of metal construction and incorporates a chromel/alumel thermocouple junction connected close to the heated filament.

Signal conditioning
The thermocouple produces a small millivolt signal proportional to the gas pressure and this must be amplified to move the indicating pointer.

Range
Thermocouple gauges are used for measuring pressures in the range from 4 mbar down to 10^{-3} mbar and have similar response times to Pirani gauges.

10.8.4 Ionisation gauges

Ionisation gauges operate by applying a voltage to electrodes surrounded by the gas whose pressure is to be measured, causing a fraction of the gas molecules to ionise. The ionisation current which flows is a measure of the pressure.

The measuring system consists of a gauge head which is essentially a thermionic-triode assembly consisting of an anode, a grid, and a filament cathode. Movement of electrons from the heated cathode causes ionisation of the gas molecules, and the ions are attracted to the anode, causing an ionisation current in the anode circuit. The gauge head is connected to the measuring instrument and this records the ionisation current, which is a measure of the gas pressure, on a calibrated scale.

Range
Static pressures from 10^{-3} mbar down to 10^{-10} mbar can be measured.

10.9 Calibration of pressure transducers

The standard method of calibrating all pressure transducers is to use a dead-weight tester with a design similar to the type illustrated in fig. 9.15. The calibration procedure is the same, but standard masses are used which have been calibrated in pressure units dependent on the area of the load ram of the tester.

Note: All dead-weight testers are accurate only if the local value of gravitational acceleration 'g' corresponds to the one under which the tester was calibrated; if it is different, corrections must be applied. Correction charts are usually supplied by the manufacturers of the dead-weight tester.

Example 10.5 A dead-weight tester is calibrated for standard gravity, 9.80665 m/s^2, but is used at sea-level on the equator where gravity is only 9.78030 m/s^2. When the indicated pressure is 10 bar, what is the true pressure?

$$\text{True pressure} = \text{indicated pressure} \times \frac{\text{local gravity}}{\text{standard gravity}}$$

$$= 10\text{bar} \times \frac{9.78030\,\text{m/s}^2}{9.80665\,\text{m/s}^2}$$

$$= 9.973\,\text{bar}$$

A secondary calibration procedure uses precision-made standard gauges (checked on a dead-weight tester before use) and a pressure comparator. A pressure comparator is a device in which the same hydraulic pressure is applied to the standard gauge and to the transducer (or gauge) under test, thus enabling the transducer to be checked relative to the standard gauge. This method is used as an industrial calibration procedure, enabling rapid checking of factory pressure gauges and transducers to a limited accuracy.

10.10 Complete measurement systems

10.10.1 Measurement of air flow in the inlet manifold of an i.c. engine

Figure 10.15 shows an arrangement suitable for measuring the air flow to an internal-combustion engine. The air flow is converted to a differential pressure by means of an orifice plate, and this small pressure differential is measured using an inclined manometer. This method is suitable for static measurements, i.e. mean air flows, but not for measuring flow fluctuations.

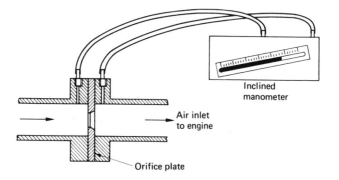

Fig. 10.15 Measurement of air flow to an i.c. engine

10.10.2 Measurement of cylinder pressure in a diesel engine

Figure 10.16 shows an arrangement for measuring the rapidly changing gas pressures within the cylinder of a diesel engine. This application requires a transducer capable of measuring maximum pressures in excess of 100 bar under extreme temperature conditions. In order to keep the transducer within its permitted temperature range, water or air cooling must be employed as shown.

A charge amplifier is used to convert the charge from the transducer into a voltage suitable for a CRO fitted with an automatic camera facility for photographing the trace of the pressure pulse.

10.10.3 Leak testing of pressurised containers

Figure 10.17 shows an arrangement for testing a pressurised container by comparing it with a leak-free identical reference unit. The same pressure is applied to both containers, which are connected to opposite sides of a differential pressure transducer operating on the strain-gauged-

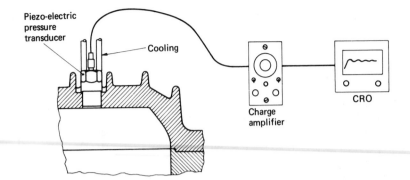

Fig. 10.16 Measurement of cylinder pressures

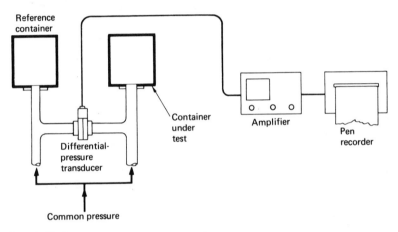

Fig. 10.17 A pressure-test arrangement

diaphragm principle. Any leaks in the container under test will be sensed as a pressure difference by the transducer. This method ensures that any pressure difference induced due to warming or cooling of the air will be insignificant, as the same effect will occur in both containers.

Exercises on chapter 10
1 Explain the difference between vented gauge pressure and sealed gauge pressure.
2 State two ways in which the sensitivity of a manometer can be increased.
3 How is low 'g' sensitivity achieved in diaphragm-type pressure gauges?

Table 10.2 Typical characteristics of pressure transducers quoted in manufacturers' specifications

Type	Pressure range	Frequency range	Excitation voltage	Output	Temperature range
U-tube	1mm to 1m water or mercury	Static measurement	—	—	For use at room temp., 20°C
Well	1mm to 1m water or mercury	Static	—	—	Room temp.
Inclined	1mm to 300mm water	Static	—	—	Room temp.
Bourdon tube	Up to 5000bar	Static	—	—	0–50°C. Higher if gauge remote
Diaphragm displacement (l.v.d.t.)	Up to 200bar	0.5kHz low press. 15kHz high press.	6V r.m.s. at 5kHz	8mV/V	−20°C to +100°C
Strain gauge:					
unbonded	Up to 700bar	10kHz low press. 50kHz high press.	10V d.c.	3.5mV/V	−10°C to +120°C
bonded	Up to 500bar	Up to 2kHz	24V d.c.	2mV/V	−30°C to +85°C
semiconductor	50mbar miniature up to 700bar normal	Up to 50kHz	10V d.c.	100mV–10V	−40°C to +150°C
vapour-deposited	Up to 2000bar	12kHz low press. 100kHz high press.	10V d.c.	2mV/V	−40°C to +120°C

Table 10.2 Typical characteristics of pressure transducers quoted in manufacturers' specifications (*continued*)

Type	Pressure range	Frequency range	Excitation voltage	Output	Temperature range
Piezo-electric	Up to 5000 bar	2 kHz low press. 150 kHz high press.	—	2000 pC/bar low press. 3 pC/bar high, 100 pC/bar typical	−150°C to +240°C
Vacuum gauge: McLeod	10 down to 10^{-3} mbar	Static	—	—	For use at room temp.
Pirani	100 down to 10^{-4} mbar	Static	—	0–10 mV, 0–600 mV aux. o/p	0–50°C
ionisation	10^{-3} down to 10^{-10} mbar	Static	—	0–10 mV	0–50°C
capsule	1000 down to 0.5 mbar	Static	—	0–1 mA from elect. operated gauge	0–50°C

4 A U-tube manometer is used to measure the difference in pressure across an orifice plate in a pipe carrying water. Determine the pressure difference when the manometer indicates a head of 0.2 m of mercury. The density of water is $1 \, Mg/m^3$ and the relative density of mercury is 13.6. [$24.72 \, kN/m^2$ or $0.247 \, bar$]

5 The differential pressure transducer shown in fig. 10.1 has the pressure $p_1 = 5 \, bar$ absolute. If the other inlet pressure port is open to atmosphere, calculate the absolute, gauge, and differential pressures sensed by the sensing element. Atmospheric pressure can be taken as 1.013 bar. [Not applicable; 3.987 bar; 3.987 bar]

6 The following table gives the results of the calibration of a 0 to 10 bar pressure gauge.

True pressure (bar)	0	2	4	6	8	10
Recorded pressure (bar)	0.05	2.05	4.00	5.95	7.93	9.9

Draw a calibration graph for the gauge under test, and determine the error on a reading of (a) 5 bar and (b) 9.4 bar as a percentage of full scale. Does the gauge meet its stated accuracy of ±2% f.s.d.? [0%; −1%; Yes]

7 A piezo-electric transducer of sensitivity 80 pC/bar is used to measure a mean pressure. The transducer signal is fed into a charge amplifier which has the sensitivity ranges of 0.05, 0.1, 0.2, 1, 2, 5, 20, and 100 mV/pC.

The charge-amplifier signal is displayed on an oscilloscope whose sensitivity is set at 1 V/cm. If the pressure being measured is 21 bar, what charge-amplifier range would be required to obtain a trace deflection of approximately 80 mm? What would the actual trace deflection be at this setting? [5 mV/pC; 84 mm]

8 A pressure transducer has a sensitivity of 1.5 mV/bar and an output impedance of 200 Ω. If this is connected to a galvanometer of resistance 50 Ω having a sensitivity of 10 mm/μA, calculate the pressure being measured if the galvanometer spot deflects 50 mm on the ultra-violet-sensitive paper. [0.83 bar]

9 It is proposed to monitor the water level in a reservoir by means of a submerged pressure transducer. Suggest a possible method.

10 Propose a method for measuring the dynamic-pressure variations in the exhaust manifold of an internal-combustion engine.

11   Vibration and noise measurement

11.1 Vibration

11.1.1 Definition
Basically, vibration is oscillating motion of a particle or body about a fixed reference point. Such motion may be simple harmonic (sinusoidal) or complex (non-sinusoidal). It can also occur in various modes – such as bending or translational modes – and, since the vibration can occur in more than one mode simultaneously, its analysis is a very exacting and complex area of engineering.

11.1.2 Units of vibration
The units of vibration depend on the vibrational parameter, as follows:

a) acceleration, measured in 'g' or m/s^2;
b) velocity, measured in m/s;
c) displacement, measured in m.

11.1.3 Some effects of vibration
Vibration can cause damage to structures and machine sub-assemblies, resulting in mis-operation, excessive wear, or even fatigue failure.

    Vibration may have adverse effects on human beings. The primary effects are task-performance interference; motion sickness; breathing and speech disturbance; and a hand-tool disease known as 'white finger', where the nerves in the fingers are permanently damaged, resulting in loss of touch sensitivity.

11.1.4 Characteristics of vibration
Vibration may be characterised by

a) the frequency in Hz;
b) the amplitude of the measured parameter, which may be displacement, velocity, or acceleration. This is normally referred to as the *vibration* amplitude when expressed in units, but vibration level when expressed in decibels.

11.1.5 Decibel notation applied to vibration measurement
Because of the wide range of vibration amplitudes found in engineering, it is convenient to express the measured amplitude in decibels with reference to a fixed value. Reference values which are internationally accepted are as follows:

a) for velocity, the reference is 10^{-3} m/s;
b) for acceleration, the reference is 10^{-5} m/s².

Thus if the measured amplitude is A_1 and the reference amplitude is A_0, the vibration level expressed in decibels is

$$\text{vibration level} = 20\log_{10}\frac{A_1}{A_0}\text{dB} \qquad 11.1$$

Example 11.1 If the measured vibrational acceleration amplitude of a body is $2g$, express this in dB ref. 10^{-5} m/s².

Using equation 11.1 and if $g = 9.81$ m/s²,

$$\begin{aligned}\text{accelerational} \atop \text{vibration level} &= 20\log_{10}\frac{A_1}{A_0}\\[2mm]
&= 20\log_{10}\frac{2 \times 9.81\,\text{m/s}^2}{10^{-5}\,\text{m/s}^2}\\[2mm]
&= 125.8\,\text{dB}\end{aligned}$$

In practice, this may be rounded off to 126 dB.

Example 11.2 If a mechanism has a vibrational velocity amplitude of 3.123 m/s, express this in dB ref. 10^{-3} m/s.

Using equation 11.1,

$$\begin{aligned}\text{velocity vibration level} &= 20\log_{10}\frac{A_1}{A_0}\\[2mm]
&= 20\log_{10}\frac{3.123\,\text{m/s}}{10^{-3}\,\text{m/s}}\\[2mm]
&= 69.89\,\text{dB}\end{aligned}$$

In practice this would be rounded off to 70 dB.

11.1.6 Relationship between the vibration parameters

Assuming that the vibration is simple harmonic motion, then

$$\text{displacement } x = A \sin \omega t$$

$$\text{velocity} \qquad v = \frac{dx}{dt} = A\omega \cos \omega t$$

$$\text{acceleration } a = \frac{dv}{dt} = -A\omega^2 \sin \omega t$$

where $\omega = 2\pi f$ rad/s

and f = frequency of vibration in Hz

200

Note that the frequencies are the same in each case, although there is a phase shift. The amplitudes of the above parameters are thus

displacement amplitude = A

velocity amplitude = $A\omega$

acceleration amplitude = $A\omega^2$

hence   velocity amplitude = $\dfrac{\text{acceleration amplitude}}{\omega}$ 11.2

and   displacement amplitude = $\dfrac{\text{velocity amplitude}}{\omega}$ 11.3

$= \dfrac{\text{acceleration amplitude}}{\omega^2}$ 11.4

Example 11.3   If the vibrational acceleration amplitude of a body is $20g$ at 10 Hz, determine (a) the velocity amplitude and vibration level; (b) the displacement amplitude.

a) Using equation 11.2 and if $g = 9.81\,\text{m/s}^2$,

acceleration amplitude $= 20 \times 9.81\,\text{m/s}^2$

$\therefore$ velocity amplitude $= \dfrac{\text{acceleration amplitude}}{\omega}$

$= \dfrac{20 \times 9.81\,\text{m/s}^2}{(2\pi \times 10)\,\text{rad/s}}$

$= 3.122\,\text{m/s}$

Using equation 11.1,

velocity vibration level $= 20\log_{10}\dfrac{A_1}{A_0}$

$= 20\log_{10}\dfrac{3.122\,\text{m/s}}{10^{-3}\,\text{m/s}}$

$= 69.88\,\text{dB ref. } 10^{-3}\,\text{m/s}$

b) Using equation 11.3,

displacement amplitude $= \dfrac{\text{velocity amplitude}}{\omega}$

$= \dfrac{3.122\,\text{m/s}}{20\pi\,\text{rad/s}}$

$= 0.05\,\text{m}$

11.1.7 Which parameter?

The choice of the best parameter to be measured depends on a number of factors, including

a) the type and size of the transducer available,
b) the mass of the vibrating structure, and
c) the frequency and amplitude characteristics of the vibration.

If the velocity, acceleration, and displacement amplitudes are measured at various frequencies, the resulting graphs of amplitude versus frequency are referred to as the vibration spectra, and the shape of the graphs are referred to as the spectral shapes.

With instrumentation based on accelerometer transducers and integrator amplifiers, the user is free to choose between acceleration, velocity and displacement as the measurement parameter. The typical vibration spectra shown in fig. 11.1 are displays of the three parameters of a machine's vibration. Although they each have different average slopes their peaks occur at the same frequencies. In the example shown the amplitude range required to display the velocity spectrum is the smallest and thus occupies the least dynamic range. In addition it means that all the frequency components on this curve need a smaller relative change before they begin to influence the overall vibration level. The low frequency acceleration and high frequency displacement components of the spectra shown in fig. 11.1 need to exhibit much larger changes before they influence the overall vibration level. In general it is therefore advisable to display in turn each of the three parameters and choose the one which has the flattest spectrum. This will enable one to detect machine faults, which produce an increase in vibration level, at an early stage. In practice the velocity–frequency spectra of many industrial machines are shaped this way, i.e. they are quite flat over a wide range of frequencies. Since it

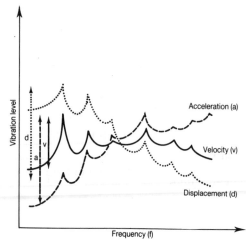

Fig. 11.1 Vibrational spectra of a machine

is also a measure of vibrational energy present, the velocity parameter is the one in most common use and its use is supported by American and European standardisation organisations.

11.1.8 Effect of the transducer on the vibrating structure

In general, the larger the mass of the vibration transducer, the greater its sensitivity. Unfortunately, the addition of the transducer's mass (m_1) to the mass (m_0) of the vibrating structure changes the resonant frequency of the vibrating system as follows:

$$\frac{f_1}{f_0} = \sqrt{\frac{m_0}{m_0 + m_1}} \qquad\qquad 11.5$$

where f_1 = resonant frequency of the structure with the mass added

and f_0 = resonant frequency of the structure before the transducer is added

Example 11.4 A transducer having a mass of 0.05 kg is attached to a vibrating structure which has a mass of 50 kg. If the resonant frequency of the structure alone is 10 Hz, determine the new resonant frequency when the mass of the transducer is added.

Using equation 11.5,

$$\frac{f_1}{f_0} = \sqrt{\frac{m_0}{m_0 + m_1}}$$

$$\therefore \quad f_1 = \sqrt{\frac{m_0}{m_0 + m_1}} \times f_0$$

$$= \sqrt{\frac{50\,\text{kg}}{50.05\,\text{kg}}} \times 10\,\text{Hz}$$

$$= 9.995\,\text{Hz}$$

11.2 Vibration-measuring devices

Vibration-measuring devices may be grouped into

a) mechanical or non-electrical transducer types;
b) electrical transducer types.

The division is not always clear, since both groups may include hybrids which use both mechanical and electrical principles.

11.2.1 Mechanical or non-electrical transducer type

Three examples of how vibration may be measured using non-electrical devices are as follows.

a) The stroboscope method

The fixed pointer or stud, shown in fig. 11.2, is attached to the vibrating surface and is used to give an indication of the displacement only. By using the light of a stroboscope to 'freeze' or 'slowly move' the stud, quite high-frequency small-amplitude vibrations may be measured. The typical upper range of frequency is quoted at 500 Hz for direct measurement.

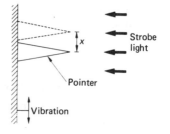

Fig. 11.2 Stroboscope method of vibration measurement

b) The reed vibrometer

The variable-length reed vibrometer shown in fig. 11.3 is used to measure the main frequency component of the vibration. In practice the length l is adjusted until the maximum reed vibration occurs, when its resonant frequency is the same as the frequency of the vibrating mechanism or structure. The length l is calibrated directly in Hz. A small mass may be added to the cantilever if the vibrometer is to be used for very-low-frequency investigation, but the scale readings would then need to be corrected for the additional mass. The range of measurement is quoted as 5 Hz to 10000 Hz.

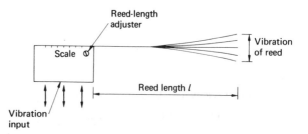

Fig. 11.3 The reed vibrometer

c) The seismic-mass transducer

The word 'seismic' comes from the Greek for 'shaking' and is used to describe the absolute-motion devices such as the seismograph which measures the vibration caused by earth tremors or quakes.

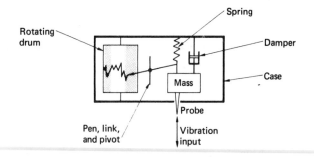

Fig. 11.4 A seismic-mass measuring device

The seismic-mass transducer shown in fig. 11.4 consists of a mass, suitably damped and spring-supported, connected to a pen which is used to trace the vibration waveform on to moving paper. The device is suitable for low-frequency vibrational-displacement measurement. The ranges of measurement are quoted as $0.01g$ to $50g$ acceleration range and $0.1\,\text{mm}$ to $15\,\text{mm}$ displacement range.

The obvious disadvantage of these three devices is that they do not permit measurement remote from the vibrating source, and they are therefore most unsuitable if a detailed analysis of the vibration is required.

11.2.2 Electrical vibrational transducers
Electrical vibration transducers are classified into two groups as follows:

a) relative-motion transducers, where the vibration may be measured relative to a stationary 'fixed' reference;
b) absolute-motion transducers, which employ seismic masses as the sensing elements.

a) Relative-motion transducers
Some examples of devices suitable for relative-vibration measurement are shown in figs 11.5(a) to (d), the principles of which are examined in greater detail in chapter 3. Their main disadvantage is that they require a fixed reference point, although there is no direct mechanical linkage between the vibrating surface and the transducing element. The distinction between displacement and vibrational amplitude is not clear, but in general any displacement-measuring device may be used for vibrational-amplitude measurement.

b) Absolute-motion transducers
In applications where it is impossible to establish a fixed reference for vibration measurement, e.g. in a moving vehicle, the transducers use the response of a seismic-mass–spring–damper system. The seismic mass is

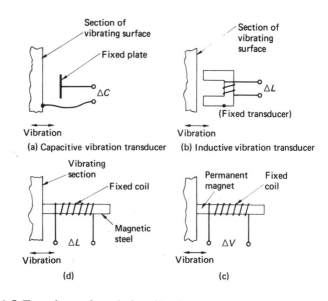

Fig. 11.5 Transducers for relative vibration measurement

held by a stiff spring, and motion of the mass within the case is damped by a viscous fluid. When the transducer casing is attached to the vibrating surface, the device may indicate displacement, velocity, or acceleration – depending on the frequency range which is used.

Figure 11.6(a) shows a vibration transducer employing the electromagnetic principle to measure velocity. The permanent magnet, acting as the seismic mass moving in response to a vibration, slides along a machined steel shaft. As it does so, it generates a voltage in the coil proportional to the rate at which the flux linkage changes, i.e. proportional to velocity. One advantage of this type of device is that its output voltage is often large enough for a signal conditioner not to be required.

Figure 11.6(b) shows the cross-section of a piezo-electric transducer in which a pre-loaded mass is held in contact with quartz discs by a stiff spring. Under the influence of acceleration, the seismic mass will exert a force on the discs in the direction of the axis of the transducer. The principles of piezo-electric transducers have been discussed in chapter 3; therefore it is only necessary to summarise the main points at this stage.

The sensitivity may be quoted as the charge sensitivity in pC/g, or as the voltage sensitivity in mV/g. The device, acting as a capacitor being charged, has a very high output impedance and is thus very sensitive to loading effects by the signal conditioner or display device. These problems are overcome by using charge amplifiers which provide an output voltage from a charge input and have been developed to have input resistances as high as 10^{14} ohms to overcome the discharge of the crystal capacitance by

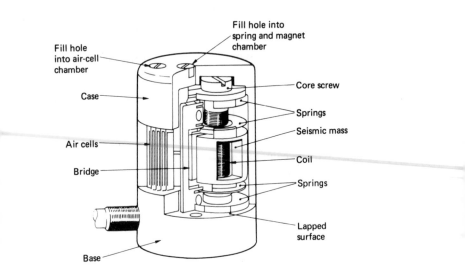

(a) Electromagnetic seismic-mass velocity transducer

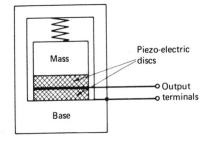

(b) Piezo-electric accelerometer

Fig. 11.6 Transducers for absolute vibration measurement

the signal conditioner. This discharge is particularly troublesome at zero or very low frequencies.

Example 11.5 The following details are extracted from a vibration transducer's specification:

Type   electromagnetic velocity transducer
Frequency range   20 to 1000 Hz
Amplitude range   5 mm peak-to-peak max.
Acceleration range   0.1 to 30g peak
Sensitivity into 10000 Ω resistive load   4.88 ± 0.2 V per m/s
Mass   170 grams
Damped natural frequency   5 Hz
Coil resistance   600 Ω
Transverse sensitivity   2% max.

a) Calculate the output voltage, correctly loaded, when sensing the maximum amplitude at the minimum frequency of the range quoted.
b) Determine the velocity sensed when the output voltage is 0.5 V peak at 100 Hz, and determine the corresponding displacement.

a) Maximum amplitude = 5 mm

   Minimum frequency = 20 Hz

Using equation 11.3,

$$\text{velocity} = \text{displacement} \times \omega$$
$$= 5 \times 10^{-3}\,\text{m} \times 2\pi \times 20\,\text{Hz}$$
$$= 0.2\pi\,\text{m/s}$$

$\therefore$ output voltage $= 0.2\pi\,\text{m/s} \times 4.88\,\text{V/(m/s)}$
$$= 3.07\,\text{V}$$

b) Output voltage $= 0.5\,\text{V peak}$

$\therefore$ peak velocity $= \dfrac{0.5\,\text{V}}{4.88\,\text{V/(m/s)}}$
$$= 0.102\,\text{m/s}$$

Using equation 11.3,

peak displacement $= \dfrac{\text{velocity}}{\omega}$

$$= \dfrac{0.102\,\text{m/s}}{2\pi \times 100\,\text{Hz}}$$
$$= 0.000\,16\,\text{m} = 0.16\,\text{mm}$$

Note: The transducer has a low natural frequency and is operated above that natural frequency.

Example 11.6 The following details are extracted from a vibration transducer's specification:

Type   quartz accelerometer
Frequency range   d.c. to 15 kHz
Range   ±3500g
Sensitivity   5 pC/g
Damped natural frequency   22 kHz
Mass   0.05 kg
Resolution   0.002g
Transverse sensitivity   2% maximum

a) Calculate the charge output at a vibrational acceleration of 0.21 g.
b) Determine the maximum charge output when the device is subjected to a transverse vibration of 1000g.

a) Charge output = acceleration × transducer sensitivity

$$= 0.21g \times 5pC/g$$

$$= 1.05pC$$

b) Main-axis equivalent input $= \dfrac{2}{100} \times$ transverse input

$$= \dfrac{2}{100} \times 1000g$$

$$= 20g$$

Charge output = input acceleration × transducer sensitivity

$$= 20g \times 5pC/g$$

$$= 100pC$$

Note: The transducer has a high natural frequency and is operated below that natural frequency.

Example 11.7 A quartz accelerometer and charge amplifier are used to measure the vibration of a machine. Given

accelerometer sensitivity = $5pC/g$

charge amplifier sensitivity = $50mV/pC$

output-voltage amplitude corresponding to peak acceleration = $2V$

calculate the vibrational acceleration of the machine.

System sensitivity K_s = transducer sensitivity × charge-amplifier sensitivity

$$= 5pC/g \times 50mV/pC$$

$$= 250mV/g$$

Measured acceleration $= \dfrac{\text{measured voltage}}{K_s}$

$$= \dfrac{2V}{250mV/g} = 8g$$

11.2.3 Comparison of vibration-measuring systems
Table 11.1 compares some features of complete vibration-measuring systems and reveals that the accelerometer system, although the most expensive, covers the widest range of frequencies and vibration levels.

Table 11.1 Comparison of vibration-measuring systems

Transducer	Parameter	Signal conditioner	Frequency range	Remarks
Capacitive Inductive	Displacement	Amplitude modulation with bridge circuits	$0-0.1f_c$ (f_c = carrier frequency)	Usually relative displacement only.
Electro-magnetic	Velocity	May need an amplifier	15 to 1000 Hz	Poor low-frequency response
Piezo-electric	Acceleration	Charge amplifier	$0-0.3f_n$ (f_n, the natural frequency, is typically 22 kHz)	Wide range of measurement. Typical ±10000g

11.3 Complete vibration-measuring systems

Examples of three applications of vibration-measuring systems are illustrated in figs 11.7 to 11.9.

An aircraft-engine vibration-measuring system is shown in fig. 11.7. The system uses four indicators and a switching device which permits the output signal of each transducer to be read individually. Four signal lamps indicate if the vibration levels exceed a pre-set alarm level.

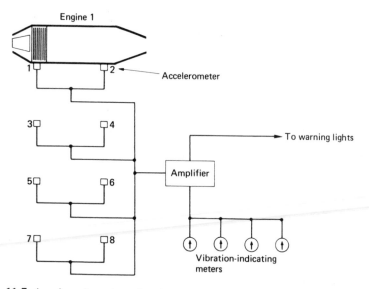

Fig. 11.7 An aircraft-engine vibration-measuring system

Figure 11.8 shows how capacitive (or inductive) transducers may be used to measure the axial vibration of a diesel engine. The system uses an amplitude-modulated system made up of an a.c.-excited capacitive bridge circuit, an amplifier, and a demodulator unit. Details of the characteristics of these devices are given in chapter 4.

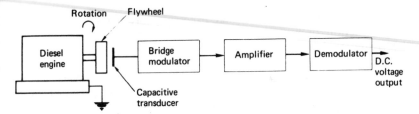

Fig. 11.8 Measurement of axial vibration of a diesel engine

A comprehensive vibration-measuring and analysing system is shown in fig. 11.9. The signal from the accelerometer and charge amplifier is frequency-analysed by a spectrum analyser. This unit gives two analogue outputs. One output provides the level recorder with an input signal for subsequent analysis, and the other output is converted into digital information for use with a computer.

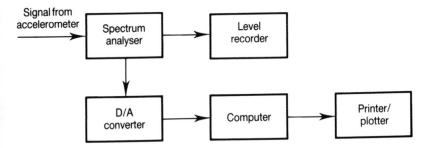

Fig. 11.9 A comprehensive vibration-measuring system

11.4 Noise

11.4.1 Definition
Noise may be defined as unwanted sound. Sound may be described as a succession of slight rapid variations in air pressure which are detectable by a hearing mechanism.

11.4.2 Some effects of noise
a) People do not, in general, like excessive noise – it causes irritation, distraction, or annoyance.
b) If the level of the noise is high enough, it may cause noise-induced hearing loss.
c) Noise interferes with communication between people.
d) By masking audible warning signals, noise reduces safety in the workshop.
e) There is some evidence that noise reduces man's working efficiency.

11.5 Characteristics of noise
Noise is characterised by

a) its intensity in watts/m^2 or its pressure in microbars;
b) its frequency in Hz;
c) the nature of the noise – it may be impulsive or continuous;
d) its loudness or loudness level;
e) its noise rating – a statistically determined 'annoyance rating' related to sound pressure level and frequency.

Characteristics (a), (b), and (c) are measured objectively by instruments and do not rely on the subjective assessment of an observer; whereas (d) and (e) are based on subjective assessment by human beings and may be determined using the measured results and sets of standard tables or graphs. The subjective reaction to noise is beyond the scope of this book, and our treatment will therefore be confined to the measurement of sound pressure levels and the basic instruments and techniques that are available for that operation.

11.5.1 Intensity and pressure of a sound wave
The intensity of the sound wave is the power per unit area, and the relationship between the intensity and its pressure is given by

$$I = \frac{p^2}{\rho_0 v} \qquad\qquad 11.6$$

where p = pressure of the sound wave (bar)

ρ_0 = density of the medium (kg/m^3)

v = velocity of the sound wave (m/s)

and I = intensity of the sound wave (W/m^2)

Hence the intensity is proportional to (pressure)^2

or $\quad I \propto p^2$ 11.7

11.5.2 Decibel notation applied to noise measurement

The human audible range of sound pressure, usually measured at 1 kHz, extends from 0.0002μbar at the 'threshold of hearing' to 1 mbar at the 'threshold of pain', an increase of 5×10^6. Because of this considerable range, it is more convenient to express the magnitude of the pressure in logarithm form, using the decibel (dB).

In appendix C it is shown how the decibel is used to express the ratio of two power variables. In sound or noise measurement, the power variable is the intensity I in W/m^2;

hence $\quad \mathrm{dB} = 10\log_{10}\dfrac{I_1}{I_0}$ 11.8

where $\quad I_0 = 10^{-12}$ W/m^2, the intensity at the threshold of hearing

and $\quad I_1 = $ the intensity of the sound being measured

Substituting from equation 11.7,

$$\mathrm{dB} = 10\log_{10}\left(\frac{p_1}{p_0}\right)^2$$

$$= 20\log_{10}\left(\frac{p_1}{p_0}\right)$$

where $\quad p_0 = 0.0002\mu$bar, the pressure at the threshold of hearing

and $\quad p_1 = $ the pressure of the sound being measured

The quantity on the left-hand side of this equation is referred to as the *sound pressure level* (s.p.l.), and most sound-measuring devices are calibrated to measure this quantity. Thus

$$\text{s.p.l.} = 20\log_{10}\frac{p_1}{p_0} \quad \text{in dB ref } 0.0002\mu\text{bar}$$ 11.9

Example 11.8 If the pressure of the sound measured at 1 m from a steel riveter is 100μbar, calculate the corresponding level in decibels ref. 0.0002μbar.

Using equation 11.9,

$$\text{s.p.l.} = 20\log_{10}\left(\frac{p_1}{p_0}\right)$$

$$= 20\log_{10}\left(\frac{100\mu\text{bar}}{0.0002\mu\text{bar}}\right) = 113.98\,\text{dB}$$

11.5.3 Variation of intensity with distance

As the distance of measurement from the noise source increases, the intensity of the sound decreases.

The simplest source of sound is the point source, where the sound is radiated equally in all directions from an apparent centre. For a point source, the relationship between distance d and intensity I is given by:

$$I \propto \frac{1}{d^2}$$

11.10

i.e. the relationship follows an inverse-square law.

In practice, industrial noise generators are rarely point sources, their size and shape being such that they are more usually rectangular or cylindrical sources. However, as a general rule of thumb, provided the distance from the source to the measuring point is greater than five times the maximum dimension of the source, the inverse-square-law relationship may be applied.

Example 11.9 The sound pressure level measured at 10m from a machine is 84 dB. Determine the sound pressure level at distances of (a) 20m and (b) 90m, assuming the inverse-square-law relationship exists between intensity and distance.

Using equation 11.10,

$$I \propto \frac{1}{d^2}$$

therefore if d is doubled, I is quartered.

a) At 10m, $\quad$ s.p.l. $= 10\log_{10}\left(\dfrac{I_1}{I_0}\right) = 84\,\text{dB}$

$\quad$ At 20m, $\quad$ s.p.l. $= 10\log_{10}\left(\dfrac{I_1/4}{I_0}\right)$

$$= 10\log_{10}\left(\frac{I_1}{I_0}\right) + 10\log_{10}\left(\frac{1}{4}\right)$$

$$= 84 - 6$$

$$= 78\,\text{dB}$$

Note that doubling the distance reduces the sound pressure level by 6 dB.

b) At 90m, $\quad$ s.p.l. $= 10\log_{10}\left(\dfrac{I_1}{I_0}\right) + 10\log_{10}\left(\dfrac{100}{8100}\right)$

$$= 84 - 19$$

$$= 65\,\text{dB}$$

11.5.4 Addition of sound pressure levels

Consider two sound pressure levels, XdB and YdB having intensities I_1 and I_2, which are added together.

Let $X\,\mathrm{dB} = 10\log_{10}\left(\dfrac{I_1}{I_0}\right)$ and $Y\,\mathrm{dB} = 10\log_{10}\left(\dfrac{I_2}{I_0}\right)$

$\therefore\quad I_1 = I_0\,\mathrm{antilog}\left(\dfrac{X}{10}\right)$ and $I_2 = I_0\,\mathrm{antilog}\left(\dfrac{Y}{10}\right)$

Resultant intensity $= I_1 + I_2$

$\therefore\quad$ resultant s.p.l. $= 10\log_{10}\left(\dfrac{I_1 + I_2}{I_0}\right)$

$$= 10\log_{10}\left(\dfrac{I_0}{I_0}\right)\left[\,\mathrm{antilog}\left(\dfrac{X}{10}\right) + \mathrm{antilog}\left(\dfrac{Y}{10}\right)\right]$$

$$= 10\log_{10}\left[\,\mathrm{antilog}\left(\dfrac{X}{10}\right) + \mathrm{antilog}\left(\dfrac{Y}{10}\right)\right]$$

Similarly, if there are several sources, with s.p.l.'s A, B, C, etc.,

$$\text{total dB} = 10\log_{10}\left[\,\mathrm{antilog}\left(\dfrac{A}{10}\right) + \mathrm{antilog}\left(\dfrac{B}{10}\right)\right.$$

$$\left. + \mathrm{antilog}\left(\dfrac{C}{10}\right) + \text{etc.}\right] \qquad\qquad 11.11$$

Thus, if $X = Y$, i.e. the s.p.l. is doubled, using equation 11.11 gives

$$\text{total s.p.l.} = 10\log_{10}\left[2\,\mathrm{antilog}\left(\dfrac{X}{10}\right)\right]$$

$$= 10\left[0.301 + \left(\dfrac{X}{10}\right)\right]$$

$$= X + 3\,\mathrm{dB}$$

i.e. doubling the sound pressure level is equivalent to adding 3 dB. Thus $90\,\mathrm{dB} + 90\,\mathrm{dB} = 93\,\mathrm{dB}$.

Example 11.10 Two machines produce individual sound pressure levels of 83 and 87 dB. Calculate their combined sound pressure levels.

Using equation 11.11,

$$\text{combined s.p.l.} = 10\log_{10}\left[\,\mathrm{antilog}\left(\dfrac{83}{10}\right) + \mathrm{antilog}\left(\dfrac{87}{10}\right)\right]$$

$$= 10\log_{10}(1.995 \times 10^8 + 5 \times 10^8)$$

$$= 88.46\,\mathrm{dB}$$

215

Example 11.11 A machine is working in a noisy environment. If the background noise when the machine is inoperative is 70dB and the combined s.p.l. of the background noise and the operating machine is 82 dB, determine the sound pressure level of the machine alone.

Using equation 11.11,

$$\text{s.p.l.} = 10\log_{10}\left[\text{antilog}\left(\frac{X}{10}\right) + \text{antilog}\left(\frac{Y}{10}\right)\right]$$

$$\therefore \quad 82\,\text{dB} = 10\log_{10}\left[\text{antilog}\left(\frac{X}{10}\right) + \text{antilog}\left(\frac{70}{10}\right)\right]$$

where X dB is the sound pressure level of the machine,

$$\therefore \quad \text{antilog}\left(\frac{X}{10}\right) = \text{antilog}\left(\frac{82}{10}\right) - \text{antilog}\left(\frac{70}{10}\right)$$

Taking $\log_{10}$ of both sides,

$$X = 10\log_{10}\left(\text{antilog } 8.2 - \text{antilog } 7\right)$$
$$= 81.7\,\text{dB}$$

i.e. the sound pressure level of the machine is 81.7 dB.

11.6 Typical noise-measuring system

The block diagram of a typical basic noise-measuring system is shown in fig. 11.10. This includes a microphone, an electronic amplifier with frequency-weighting networks, and a meter or recorder calibrated in decibels.

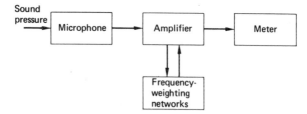

Fig. 11.10 A basic noise-measuring system

The frequency response of the unweighted system would normally be flat over the whole audible range between 20Hz and 20kHz, although some commercially available sound-level meters have an upper frequency-measuring range of approximately 30kHz.

The instrument may be weighted to provide a response similar to that of the human ear, so that an indication of subjective assessments may be made as well as objective measurements. Figure 11.11 shows two forms of

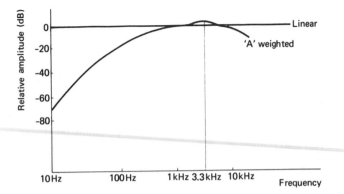

Fig. 11.11 Linear and 'A'-weighted response of sound-level meters

response which are available on some commercially available instruments:

a) *Linear response* This is the normal flat response of the microphone and electronic amplifier between, typically, 20 Hz to 31.5 kHz. Results are quoted in dB or dB 'linear'.

b) *'A'-weighted response* Electrical filters are switched into the system so that the instrument's frequency response is similar to that of the human ear. These filters reduce or attenuate the sound at the lower frequencies and increase or enhance the sound around 3 kHz. Note that the response is flat around 1 kHz, the frequency used as a reference in many subjective auditory tests. Results are quoted in dB(A) or dB'A'.

11.7 Microphones

Microphones are transducers which convert sound pressure variations into a voltage. The main factors in the selection of a microphone are frequency response, sensitivity, dynamic range, and linearity, and it is interesting to compare these features of three microphone types, as in Table 11.2.

Table 11.2 Microphone comparison (A = excellent, B = acceptable, C = unsatisfactory)

Feature	Electromagnetic	Piezo-electric	Capacitor
Frequency response	C	B	A
Sensitivity	A	C	B
Dynamic range	B	A	A
Linearity	C	B	A

These comparisons show that the capacitor microphone is the most useful, having a good frequency response, excellent linearity, and large dynamic-pressure measuring range. A typical construction of this type of microphone is illustrated in fig. 11.12(a). The microphone cartridge consists of a thin metallic diaphragm in close proximity to a rigid backplate. The diaphragm and backplate thus form the plates of a capacitor, and the variations in air pressure due to sound waves cause the diaphragm to be displaced, thereby changing the capacitance C. The charge to the capacitor is maintained at a constant level, resulting in output-voltage changes which are proportional to the pressure input.

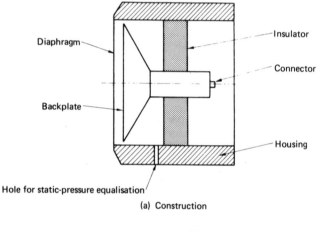

(a) Construction

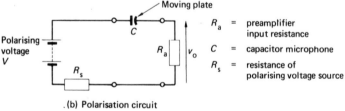

.(b) Polarisation circuit

Fig. 11.12 The capacitor microphone

The simplified circuit diagram shown in fig. 11.12(b) indicates how the output voltage is obtained from the series combination of C and R_s, the resistance of the polarising voltage source.

11.8 Frequency analysis of vibration and noise
Control of vibration and noise depends largely on the determination of the main frequency components of, quite often, complex waveforms. Spectral shapes, i.e. graphs of vibrational or noise level in dB against

frequency, enable the engineer to determine what measures are required to reduce the level of vibration and noise; hence frequency analysis is an increasingly important area of engineering. Vibration and noise may be frequency-analysed by performing the measurements in selected frequency bands. These bands may be either 1 octave, ½ octave, or ⅓ octave, where an octave corresponds to a doubling or halving of frequency. Thus 2 kHz is one octave away from 1 kHz; similarly 500 Hz is one octave away from 1 kHz.

Internationally agreed preferred octave bands of frequencies are used extensively for both vibration and noise analysis, and these are as follows:

Preferred
centre
frequency (Hz)   31.5 63 125 250 500 1 k 2k 4k 8k 16k 31.5 k

Example 11.12 A 1 octave analysis of machinery noise produced the following results.

Centre frequency (Hz)	31.5	63	125	250	500	1k	2k	4k	8k	16k	31.5k
S.P.L. (dB)	87	83	84	81	80	81	72	70	66	60	52

Determine the overall sound pressure levels in dB.

Using equation 11.11 to add together the sound pressure levels of each octave band,

$$s.p.l. = 10\log_{10} [\text{antilog } 8.7 + \text{antilog } 8.3 + \text{antilog } 8.4 +$$
$$\text{antilog } 8.1 + \text{antilog } 8.0 + \text{antilog } 8.1 +$$
$$\text{antilog } 7.2 + \text{antilog } 7.0 + \text{antilog } 6.6 +$$
$$\text{antilog } 6.0 + \text{antilog } 5.2] \, dB$$
$$= 91.2 \, dB$$

11.9 The Noise at Work Regulations
The Noise at Work Regulations 1989 came into force in January 1990. These new regulations are based on the requirements of European Community Directive 86/188/EEC which is designed to reduce the damage to hearing caused by loud noise in the workplace.

Action required by the regulations depends upon the level of noise exposure. At 85 dB(A) over a working day, employers have to make noise assessments, give employees information and issue personal ear protection on request. At 90 dB(A) employers must reduce noise where reasonably practicable, designate ear protection zones and ensure employees have and wear ear protectors whenever they have to work above that level. Employees must co-operate by wearing ear protectors above 90 dB(A). Machine makers and suppliers have to provide adequate information with machines which are likely to produce noise levels in excess of 85 dB(A).

Exercises on chapter 11

1 Express the following in decibels: $20g$ acceleration, ref. 10^{-5}m/s^2; 10m/s velocity, ref. 10^{-3}m/s. [145.85 dB; 80 dB]

2 If a vibration level is measured as $8g$ at 5Hz, express this in dB ref. 10^{-5}m/s^2 and calculate the corresponding velocity in dB ref. 10^{-3}m/s. [138 dB; 67.95 dB]

3 A transducer having a mass of 0.07kg is attached to a vibrating mechanism. If mounting the transducer changes the resonant frequency of the mechanism (with transducer) by 1%, determine the mass of the mechanism alone. [3.5 kg]

4 If a transducer having the specification shown in example 11.5 is electrically connected to a meter having an input resistance of $8\text{k}\Omega$, respecify the output-voltage sensitivity of the device. [4.812 V/(m/s)]

5 An accelerometer and a charge amplifier are used to measure vibration levels. If the transducer's sensitivity is $7\text{pC}/g$ and the charge amplifier's sensitivity is 100mV/pC, determine the output voltage of the system for an input acceleration of $3g$. [2.1 V]

6 Outline the main advantages and disadvantages of the following types of microphones for sound measurement: (a) capacitor, (b) piezo-electric, (c) dynamic (moving-coil).

7 Discuss the main difference between noise measurement on the linear and 'A'-weighted networks of a noise-level meter.

8 An octave analysis of the noise output of a machine produced the following results:

Octave centre freq. (Hz)	31.5	63	125	250	500	1000	2000	4000	8000	16000	32000
S.P.L. (dB)	70	71	84	86	87	85	77	72	64	66	68

Calculate the overall sound pressure level. [91.95 dB]

9 A machine produces a noise level of 100dB linear and 82dB(A) as measured on a noise-level meter. What do these readings tell you about the main frequency components of the noise?

10 Three machines are operating together and the noise level is measured at equal distances from each machine. The individual s.p.l.'s of the machines are 94dB, 87dB, and 91dB respectively. Determine the noise level with all machines operating together. [96.3 dB]

11 If the sound pressure level at the threshold of pain is 1mbar, determine the corresponding sound pressure level in dB. [133.9 dB]

12 The following sound pressure levels were measured for a machine operating in a noisy environment:

s.p.l. of machine + background noise = 92dB

s.p.l. of background noise = 83dB

Determine the s.p.l. of the machine alone. [91.42 dB]

13 Assuming uniform radiation find the sound power of a machine whose SPL is 88dB at 10m.

12   Temperature measurement

12.1 Definition

Temperature is defined as the degree of 'hotness' of a body. It is different in nature from the other fundamental quantities – mass, length, and time – in that it is intensive rather than extensive; i.e. if two bodies of like length are combined the total length is twice the original, but the combination of two bodies at the same temperature results in exactly the same temperature.

12.2 Units

With SI units, a 'thermodynamic' scale known as the Kelvin scale is used in which the unit of temperature is the kelvin (K). Note that the degree symbol (°) is not used.

The Kelvin scale is not dependent on the properties of measuring devices but is obtained by assigning values to two fixed points. The lowest point is 0 K or absolute zero, and is based on the theoretical minimum temperature possible for any substance. The second point is the temperature of the triple point of water, which is fixed at 273.15 K, thus setting the interval in kelvins between these two temperatures.

. The triple point of water is the temperature at which ice, water, and vapour are all in equilibrium. This occurs at a pressure of 610 Pa and can be reproduced readily to a high degree of accuracy, using a device known as a triple-point cell.

To enable accurate calibration of a wide range of temperature-measuring devices in terms of the Kelvin scale, the International Practical Temperature Scale (IPTS) has been devised. This lists eleven 'fixed points' which can be reproduced accurately. Two typical values are

a) 90.188 K – the boiling point of oxygen;
b) 1235.08 K – the freezing point of silver.

The temperature scale in everyday use is the Celsius (°C) scale (less correctly called the centigrade scale) which has 0°C for the temperature of melting ice and 100°C for the temperature of boiling water (both at normal atmospheric pressure).

The relationship between the two scales is

$$x°C = (x + 273.15) K$$

In thermodynamic relationships this is approximated to

$$x°C = (x + 273) K$$

and the Kelvin temperature is often referred to as the absolute temperature.

A third scale, which has been superseded by the Celsius scale, is the Fahrenheit (°F) scale which has 32°F for the freezing point of water and 212°F for the boiling point of water.

12.3 Measurement of temperature

Temperature measurement is particularly important in energy conservation and control, where accurate measurements are required to enable energy balances to be carried out.

Temperature cannot be measured directly, but must be measured by observing the effect which temperature variations have upon different materials. The methods of temperature measurement can be broadly divided into three groups:

a) non-electrical,
b) electrical, and
c) radiation.

12.4 Non-electrical methods

All non-electrical methods depend on the thermal-expansion properties of solids, liquids, or gases when they are exposed to temperature changes. Since these devices have no electrical connections, they are commonly used in areas where there is a risk of explosion, e.g. to provide a display of temperature for petrol-storage tanks.

12.4.1 Expansion of liquids

The liquid-in-glass thermometer is one of the most common types employing the expansion-of-liquid principle. It indicates temperature due to the different expansion rates of the liquid (commonly mercury or alcohol) and the glass container. Since the coefficient of expansion of the fluid is much greater than that of glass, a temperature change causes the fluid to expand along a calibrated capillary tube.

Liquid-in-glass thermometers are accurate and very reliable, since they have no moving mechanisms to develop faults; however they are very fragile and not always suitable for industrial use. The thermometer is often enclosed in a metal guard, but full protection can be achieved only by also enclosing the bulb in a pocket or sheath. However, the additional thermal lag considerably increases the response time of the thermometer.

Industrial liquid-expansion thermometers use a metal bulb (often stainless steel) instead of glass. This results in a robust easy-to-read thermometer which may be read remotely by connecting the bulb to a Bourdon-tube type of gauge by means of a flexible metal capillary tube as illustrated in fig. 12.1. Errors can arise in this type of thermometer, due to expansion of the capillary and Bourdon tubes if these are at a different temperature to that at which the device was calibrated. This is minimised by ensuring that the sensing-bulb volume is considerably greater than that of the capillary and Bourdon tubes.

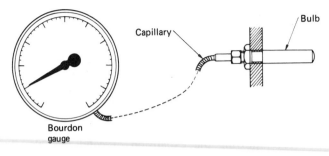

Fig. 12.1 A mercury-in-steel expansion thermometer

Signal conditioning
In the liquid-in-glass thermometer, signal conditioning is achieved by the capillary tube which converts liquid expansion into a fluid movement against a scale.

In the Bourdon-tube type of instrument the expansion can be transmitted up to 35m via a capillary to the Bourdon tube. The deflection produced is converted into a pointer motion by gears and linkages.

Example 12.1 A Bourdon-tube temperature indicator has a range of 0°C to 200°C and produces a corresponding pointer rotation of 0 to 270°. If the mechanical levers and gears have an amplification of 25, determine the sensitivity of the Bourdon tube in rad/°C.

$$\frac{\text{Angular displacement}}{\text{of Bourdon tube}} = \frac{\text{pointer movement}}{\text{mechanical amplification}}$$

$$= \frac{270°}{25}$$

$$= 10.8°, \text{ or } 0.188 \text{ radians}$$

$$\therefore \quad \frac{\text{Bourdon-tube}}{\text{sensitivity}} = \frac{\text{angular displacement}}{\text{temperature}}$$

$$= \frac{0.188 \text{ rad}}{200°C}$$

$$= 9.4 \times 10^{-4} \text{ rad/°C}$$

12.4.2 Expansion of vapours and gases
Vapour thermometers use exactly the same construction as the liquid-in-metal thermometer, i.e. a metal bulb, a flexible capillary, and a Bourdon tube. The vapour thermometer, however, uses a bulb partly filled with a volatile liquid such as methyl chloride, sulphur dioxide, or ether, the rest

of the device being filled with its vapour. A temperature increase causes more liquid to evaporate, and the resultant increase in pressure is sensed by the Bourdon tube.

Gas thermometers operate on the same principle, but an inert gas such as nitrogen is used as the operating medium. Since the gas has a lower thermal capacity than fluids, shorter response times can be achieved.

Both of these types of thermometer have non-linear scales.

12.4.3 Expansion of solids

Thermometers employing the expansion of solids use the different expansion rates of two dissimilar metals which are welded together in the form of a bimetallic strip. Application of heat results in a differential expansion which causes the strip to deflect an amount proportional to the temperature change. Typical metals used are brass or nickel as the high-expansion metal and Invar (an iron–nickel alloy) as the low-expansion metal.

This type of instrument is robust in construction and accurate, but has a relatively long response time and cannot be used remotely.

Signal conditioning

If the bimetallic strip is wound in the form of a helix, application of heat causes the helix to unwind, moving a pointer over a calibrated scale as outlined in chapter 5.

12.5 Electrical methods

There are two main electrical methods used for measuring temperature, employing either

a) self-generating transducers, i.e. thermocouples, or
b) variable-control-parameter transducers, i.e. variable-resistance transducers.

12.5.1 Thermoelectric pyrometers

Thermoelectric pyrometers are the most versatile type of temperature-measuring device available, and hence the most widely used. They all employ thermocouples as the temperature-sensitive element. The principle of the thermocouple has been outlined in section 3.13, but it consists essentially of two dissimilar conductors joined together at both ends. If one end is maintained at a constant temperature (the cold junction), the thermoelectric e.m.f. generated is a measure of the temperature of the other end (the hot junction).

The basic thermocouple construction is illustrated in fig. 12.2 and consists of two wires twisted and brazed or welded together with each wire covered in insulation, either

a) mineral (magnesium oxide) insulation for normal duty, or
b) ceramic insulation for heavy duty.

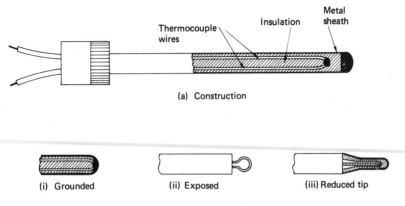

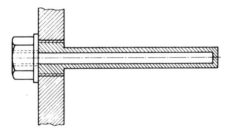

Fig. 12.2 A thermocouple construction

The whole assembly is then usually enclosed in a protective sheath which provides mechanical strength and affords protection against adverse environmental effects such as oxidation and corrosion.

When temperature sensors are to be used in very highly corrosive or abrasive media, or if the sensor needs to be removed during operation, a thermowell (or thermopocket) can be used. This is an oversheath which is permanently fitted in place via screwed or flanged connections, as illustrated in fig. 12.3.

Fig. 12.3 A thermowell cross-section

Protective sheaths and thermowells can be used only if long response times are acceptable, since they reduce the rate of heat transfer to the 'hot' junction of the thermocouple.

Example 12.2 Figure 12.4 shows the response curves for a thermocouple with and without a thermowell. By comparing the two responses, estimate the thermal time constants in each case.

225

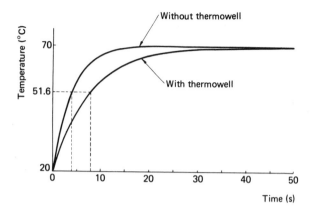

Fig. 12.4 Thermocouple responses

The time constant is defined in chapter 2 as the time to reach 63.2% of the final steady-state value.

$$\text{Total temperature change} = 70°C–20°C$$
$$= 50°C$$

and $\qquad$ 63.2% of 50°C = 31.6°C

The thermal time constant is therefore the time to reach a temperature of 51.6°C and, from fig. 12.4,

.   time constant without thermowell = 4 s

time constant with thermowell = 8 s

Thermocouples can be divided into three broad types, depending on their composition:

a) base metal,
b) rare metal, and
c) non-metallic.

a) Base-metal thermocouples
The base-metal thermocouples use combinations of the pure metal and alloys of iron, copper, and nickel and are used in the lower range of temperatures up to 1450K. They have the following advantages over rare-metal thermocouples:

i) the material of construction is cheaper, hence 3 mm diameter wire can be used instead of 0.5 mm, which results in a more robust thermo-couple.
ii) higher output voltages are obtained.

However, they are more prone to oxidation and corrosion, and their allowable temperature range is lower.

b) Rare-metal thermocouples

The rare-metal thermocouples use combinations of the pure metal and alloys of platinum and rhodium for temperatures up to 2000K and tungsten, rhenium, and molybdenum for temperatures up to 2900K.

Typical types of metal thermocouple, with their temperature ranges, are shown in Table 12.1, and some of their characteristics are shown in fig. 12.5.

Table 12.1 Thermocouple ranges (as quoted in BS 4937)

Thermocouple	Material*	Temperature range (K)
Base-metal:		
type T	Copper–constantan	3–673
type E	Chromel–constantan	3–1273
type J	Iron–constantan	63–1473
type K	Chromel–alumel	3–1643
Rare-metal		
type S	Platinum–platinum/10% rhodium	223–2033
type R	Platinum–platinum/13% rhodium	223–2033
type B	Platinum/30% rhodium–platinum/6% rhodium	273–2093

*constantan = copper/nickel; chromel = nickel/chromium;
alumel = nickel/aluminium

Example 12.3 Using fig. 12.5, determine (a) the sensitivity of the type-T thermocouple in the range 0°C to 300°C, (b) the sensitivities of the type-E and type-S thermocouples in the range 400°C to 1000°C.

a) The sensitivity is the slope of the characteristic,

$$\therefore \quad \text{sensitivity} = \frac{15.0\,\text{mV}}{300°C}$$

$$= 0.05\,\text{mV/°C}$$

b) Type-E sensitivity $= \dfrac{(76 - 28)\,\text{mV}}{(1000 - 400)°C}$

$$= \frac{48\,\text{mV}}{600°C}$$

$$= 0.08\,\text{mV/°C}$$

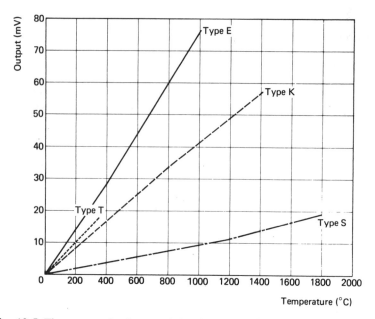

Fig. 12.5 Thermocouple characteristics (reference 0°C)

$$\text{Type-S sensitivity} = \frac{(9.5 - 3.5)\,\text{mV}}{600°C}$$

$$= 0.01\,\text{mV/°C}$$

Note the much lower sensitivity for the rare-metal type.

c) Non-metallic thermocouples

The use of non-metallic thermocouples is not very widespread; however, one novel use of them is in the power packs of the Viking 1 and 2 space explorers. Here 312 silicon–germanium thermocouples are used to convert heat from the thermal generator into about 450 watts of electrical power for driving all electrical systems on the module.

Two important considerations when using all thermoelectric pyrometers are

 i) maintaining a constant cold-junction reference temperature, and
ii) remote indication requiring the use of extension leads.

Cold-junction reference

The cold-junction temperature must be maintained at a constant value or inaccuracies will result. Three basic methods are used to achieve this:

228

a) Provision of a fixed reference temperature, such as an ice-bath, which maintains the cold junction at 0°C. This is used in laboratory measurements.
b) Provision of a stable temperature by using a thermostatically controlled cold junction held just above the maximum ambient temperature, and applying corrections for the cold junction not being 0°C.
c) By automatic cold-junction compensation, using a thermistor to sense any ambient-temperature change and cause an appropriate voltage to be added to the thermocouple voltage.

Example 12.4 A type-E thermocouple is connected to a voltmeter whose terminals are at 50°C. If the potentiometer reading is 60 mV, what is the temperature at the thermocouple hot junction?

Using fig. 12.5,

$$\text{thermocouple output at a temperature of 50°C} = 4\,\text{mV}$$

Therefore the thermocouple output based on a reference temperature of 0°C is given by $60 + 4 = 64\,\text{mV}$. This corresponds to a temperature of 850°C.

Extension leads
To overcome the necessity for using long expensive lengths of thermo-couple wire to connect the hot junction to a remote measuring instrument, extension leads are used. These leads must have thermo-electric properties similar to the thermocouple wires over their own operating temperature range, which is usually between 0°C and 100°C; e.g. copper and copper-alloy leads are used with type S, R, and B rare-metal thermocouples. Suitable extension leads are always suggested by the thermocouple manufacturers.

Signal conditioning
Output voltages generated by thermocouples are low, especially for the rare-metal type, typically 10 mV d.c. at 1000°C, necessitating the use of either very sensitive galvanometers or high-gain high-stability d.c. amplifiers.

The basic thermocouple circuit is shown in fig. 12.6(a), with the thermocouple output being connected to the indicator by means of extension leads. The indicator is a sensitive galvanometer which measures the current resulting from the thermoelectric e.m.f. Included in the circuit is a ballast resistor with a much higher resistance than the indicator. It is made from a material whose resistance does not vary with temperature, hence any resistance changes in the indicator due to temperature variations will be swamped by the ballast resistor.

The indicator scale can be calibrated in either mV or, if only one type of thermocouple is to be used, directly in °C. A correction for the cold junction not being at 0°C will have to be made.

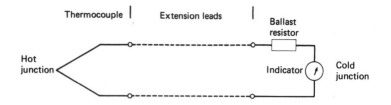

(a) The basic thermocouple circuit

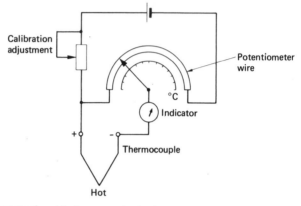

(b) The potentiometric thermocouple circuit

Fig. 12.6 Thermocouple circuits

An alternative to the above circuit uses a potentiometer to balance the thermocouple e.m.f., as illustrated in fig. 12.6(b). In this case, the sensitive galvanometer is used to detect a null-balance condition when the voltage drop down the potentiometer wire equals the thermocouple voltage. As the wiper is moved across the potentiometer wire to achieve this balance condition, a pointer is moved across a calibrated scale to give the temperature reading. Two important points to note are

i) cold-junction corrections are again necessary, and
ii) calibration of the potentiometer by means of a standard-voltage cell is necessary before each measurement.

Larger output voltages can be generated by using a series arrangement of thermocouples known as a *thermopile*. This magnifies the output by a factor n, where n is the number of thermocouples in series. This type of arrangement is particularly useful for measuring small temperature differences between the hot and cold junctions, and is often used in radiation pyrometers (discussed later in section 12.6.2).

12.5.2 Resistance thermometers
As outlined in chapter 3, resistance temperature-sensing elements can be divided into two main groups: metals and semiconductors.

a) Metallic resistance thermometers
For the measurement of lower temperatures, up to about 600°C, electrical resistance thermometers using various metals are suitable for both laboratory and industrial applications requiring

i) a high degree of accuracy, and
ii) long-term stability.

In general, the resistance of most metals over a wide temperature range is given by the quadratic relationship

$$R = R_0(1 + aT + bT^2)$$

where R = resistance at absolute temperature T

R_0 = resistance at $0\,K$

a and b = constants obtained experimentally.

However, over a limited temperature range around 0°C (273 K) the following linear relationship can be applied:

$$R = R_0(1 + \alpha\theta) \qquad\qquad 12.1$$

where α = the temperature coefficient of resistance of the material in (ohms/ohm)/°C or $°C^{-1}$

R_0 = resistance at 0°C

and θ = temperature relative to 0°C

Some typical values for α are

copper $0.0043°C^{-1}$
nickel $0.0068°C^{-1}$
platinum $0.0039°C^{-1}$

If a change in temperature from θ_1 to θ_2 is considered, equation 12.1 becomes:

$$R_2 = R_1 + R_0\alpha(\theta_2 - \theta_1)$$

Rearranging gives

$$\theta_2 = \theta_1 + \frac{(R_2 - R_1)}{\alpha R_0} \qquad\qquad 12.2$$

Example 12.5 A platinum resistance thermometer has a resistance of 138.5 Ω at 100°C. If its resistance increases to 281 Ω when it is in contact with a hot gas, determine the temperature of the gas. The resistance can be taken as 100 Ω at 0°C.

Using equation 12.2 and α for platinum as $0.0039°C^{-1}$,

$$\theta_2 = 100 + \frac{(281 - 138.5)\Omega}{0.0039°C \times 100\Omega}$$

$$= 100°C + 365.4°C$$

$$= 465.4°C$$

The temperature of the gas is therefore 465.4°C.

Metallic resistance thermometers are constructed in many forms, but the temperature-sensitive element is usually in the form of a coil of fine wire. A typical construction is shown in fig. 12.7, where the wire is wound on a grooved hollow ceramic former and covered in a protective cement. The ends of the coil are welded to stiff copper leads which are connected to terminals in the head of the transducer. In some cases this arrangement can be used directly in the medium whose temperature is being measured, allowing a fast speed of response. However, in most applications a protective metal sheath is used to provide rigidity and mechanical strength.

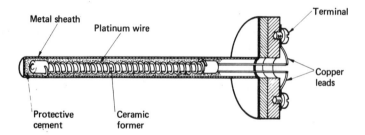

Fig. 12.7 Construction of a platinum resistance thermometer

The two most common metals used are platinum and nickel. Nickel, with its high sensitivity and reasonable price, is used in the most common base-metal resistance thermometer. Platinum, in spite of its low sensitivity and high cost, finds wide application in resistance thermometers because it has the following advantages:

i) high resistance to chemical attack and contamination,
ii) a very stable relationship over a wide temperature range, and
iii) it forms the most easily reproducible type of temperature transducer.

b) Semiconductor resistance thermometers
Semiconductor resistance elements are available with positive temperature coefficients; however, these are generally used for temperature

232

compensation and overheating protection rather than temperature measurement.

The common form of semiconductor used for temperature measurement is the thermistor, which is manufactured from oxides of copper, manganese, nickel, cobalt, and lithium, blended to give the required resistance–temperature characteristics. As outlined in chapter 3, these usually have negative temperature coefficients, i.e. NTC thermistors, and are available in many forms such as beads, wafers, and chips as illustrated in fig. 12.8.

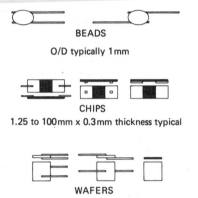

BEADS

O/D typically 1 mm

CHIPS

1.25 to 100 mm × 0.3 mm thickness typical

WAFERS

1.25 to 15 mm × 0.3 mm thickness typical

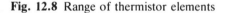

Fig. 12.8 Range of thermistor elements

The thermistor has the following advantages for temperature measurement:

i) small size (down to 1 mm diameter), enabling spot and surface temperature measurements;

ii) a large temperature coefficient, i.e. a higher sensitivity than other temperature sensors, enabling very small temperature changes to be measured;

iii) an ability to withstand electrical and mechanical stresses;

iv) a wide operating-temperature range; and

v) a wide range of resistance values.

Disadvantages are a non-linear resistance–temperature characteristic and problems of self-heating effects which necessitate the use of much lower current levels than with metallic sensors.

Table 12.2 shows typical thermistor design parameters for different types of NTC thermistor.

Dissipation constant is a measure of the self-heating effect and is given by

Table 12.2 Thermistor design parameters

Type	Temperature range (°C)	Time constant (s)	Dissipation constant (mW/°C)	Resistance at 25°C(Ω)	Resistance tolerance (%)
Bead	−55 to +300	1 to 3	<1	500 to 20M	±20 to ±1
Chip	−55 to +150	3 to 23	1 to 8	500 to 10M	±10 to ±1
Wafer	−55 to +150	3 to 23	1 to 8	3 to 20M	±10 to ±1
Flake	−55 to +200	<0.1	<1	20K to 2M	±20 to ±10

$$\frac{\text{dissipation}}{\text{constant}} = \frac{\text{change in power dissipation in the thermistor}}{\text{resultant body-temperature change}}$$

Resistance at 25°C is the nominal resistance of the thermistor at 25°C.

Resistance tolerance is the possible percentage variation in the nominal quoted resistance.

Signal conditioning

The output from the sensing element in all resistance thermometers is a change in resistance which can therefore be sensed using a bridge network. However, since the resistance change is much larger than with resistance strain gauges, special precautions must be taken in the bridge design to minimise non-linearities.

A further problem, which occurs if the bridge circuit is used remote from the sensor, is the temperature gradient across the connecting wires. Since the bridge will also sense changes in this gradient, the four-wire arrangement shown in fig. 12.9 must be used to compensate for the change.

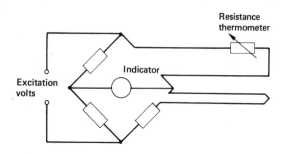

Fig. 12.9 Four-wire bridge arrangement

12.6 Radiation methods

All of the temperature-measuring methods so far discussed require contact between the sensing element and the hot body. With radiation pyrometers, however, no physical contact is required. This enables them to be used for

a) measurement of very high temperatures, and
b) measurement of the temperature of inaccessible hot bodies.

They rely on the fact that all bodies above absolute zero (0K) emit electromagnetic radiation, and the magnitude of the emission is determined by its temperature. The energy radiated is spread over a wide range of frequencies (and hence wavelengths) from around 3×10^5 MHz (10^{-3} m) up to 3×10^9 MHz (10^{-7} m). This includes the visible light range used by optical pyrometers and the 'invisible' infra-red range used by infra-red pyrometers.

For normal service, all radiation pyrometers are calibrated to read correctly when viewing a 'black body'. This is a thermodynamic concept of a body, not necessarily black in colour, which not only absorbs all energy incident upon it but also is the best possible emitter of energy. It emits radiation dependent only on its own temperature and not on reflected energy from other sources, and hence is sometimes defined as 'a body devoid of reflecting power'. A 'black body' can never be realised in practice, but one of the nearest examples occurs when viewing a furnace through a very small aperture, since any radiation emitted from this small aperture will virtually be entirely due to the furnace.

Errors can arise when pyrometers are used to measure temperatures 'out in the open', especially of hot flowing streams of molten metal; however, manufacturers supply radiation pyrometers calibrated for particular applications, which minimise these errors.

The relationship between a 'non-black body' and a 'black body' is given by emissivity, where

$$\text{emissivity } \epsilon = \frac{\text{total radiation emitted by body}}{\text{total radiation emitted by a 'black body' at the same temperature}}$$

and the power radiated per unit area from any body is given by Stefan's law as

$$P = \epsilon \sigma T^4 \qquad \qquad 12.3$$

where $P =$ total radiated power (W)

$T =$ body temperature (K)

$\sigma =$ Stefan's constant

$= 5.67 \times 10^{-8}\,\text{Wm}^{-2}\text{K}^{-4}$

and $\epsilon =$ emissivity

Example 12.6 The power radiated from a molten metal is measured and the temperature is determined to be 1500°C, assuming a surface emissivity of 0.82. If it is later found that a more accurate estimation of this emissivity is 0.75, calculate the actual temperature of the metal.

Rearranging equation 12.3 gives

$$T = \left(\frac{P}{\epsilon\sigma}\right)^{1/4}$$

$\therefore$ incorrect temperature $T_1 = \left(\dfrac{P}{\epsilon_1\sigma}\right)^{1/4}$

and actual temperature $T_{act.} = \left(\dfrac{P}{\epsilon_2\sigma}\right)^{1/4}$

$\therefore$

$$\frac{T_{act.}}{T_1} = \left(\frac{\epsilon_1}{\epsilon_2}\right)^{1/4}$$

$$T_{act.} = T_1 \times \left(\frac{\epsilon_1}{\epsilon_2}\right)^{1/4}$$

$\therefore$

$$T_{act.} = (1500 + 273)\,K \times \left(\frac{0.82}{0.75}\right)^{1/4}$$

$$= 1813\,K$$

$\therefore$ actual temperature $= 1813 - 273 = 1540°C$

BS 1041 includes a correction chart based on Stefan's law which can also be used if the emissivity of the hot body is known.

12.6.1 Optical pyrometers

Optical pyrometers use the intensity of visible radiation, with wavelengths in the range 4×10^{-7}m to 8×10^{-7}m approximately, to measure the temperature of the body. Hence they are suitable only for temperatures above about 650°C, which is the minimum temperature for the radiation to become visible.

The most common type of optical pyrometer is the disappearing-filament type, in which the intensity of visible radiation from the hot body is compared with the intensity of radiation emitted by an incandescent lamp filament. A typical arrangement is shown in fig. 12.10, where the heated filament is in the field of view of the telescope through which the heated body is viewed. The current to the lamp is varied until the filament blends with the background brilliance of the viewed body, as illustrated in fig. 12.11. This current is then a measure of the temperature of the hot body. The monochromatic screen is used so that matching is made at one particular wavelength, usually the red colour. The absorption screen absorbs some of the radiant energy from the source, enabling the heated filament to 'match' temperatures well above its own temperature.

Optical pyrometers are particularly useful for measuring the temperatures of very small objects; however they are suitable only for manual operation and not for continuous recording.

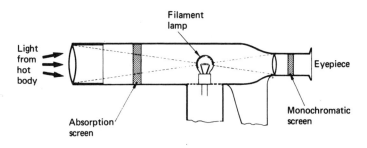

Fig. 12.10 Optical pyrometer arrangement

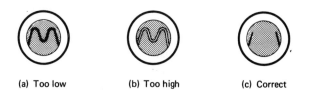

(a) Too low   (b) Too high   (c) Correct

Fig. 12.11 Filament matching for the optical pyrometer

12.6.2 Infra-red pyrometers

The infra-red pyrometer operates by focusing the infra-red radiation from the hot-body on to a temperature-sensing element such as a resistance thermometer or thermopile. The focusing system for the pyrometer is similar to a telescope arrangement, as illustrated in fig. 12.12, so that the radiation from the hot source can be focused accurately on to the temperature-sensing element.

Since the temperature is sensed using electrical sensors, continuous recording or indicating is possible. If thermocouples are used, care must be taken by the manufacturer to ensure that the cold junction is well protected from the radiated heat from the source.

Signal conditioning

A common design used for optical pyrometers uses a bridge circuit in which the filament lamp is connected in one of the arms and the balancing galvanometer is calibrated in terms of temperature.

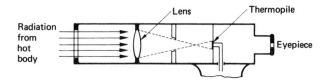

Fig. 12.12 Infra-red-pyrometer arrangement

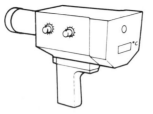

Fig. 12.13 A portable infra-red pyrometer

Since infra-red pyrometers use resistance or thermocouple temperature sensors, they use the same signal conditioning as these devices. Portable pyrometers are available with digital readout and suitable for battery or mains operation, as illustrated in fig. 12.13.

12.7 Ranges
The ranges for all the temperature-measuring devices discussed are shown in Table 12.3.

Table 12.3 Typical temperature ranges quoted in manufacturers' specifications

Type of device	Temperature range (K)	Thermal time constant
Liquid-in-glass thermometer	200–770	Around 5 s
Liquid-in-metal thermometer	230–870	10 s and higher
Vapour thermometer	250–600	10 s and higher
Gas thermometer	200–800	5 s and higher
Bimetallic	170–820	5–10 s
Thermocouple	3–2900	Less than 0.1 s for bare-metal miniature types
Resistance thermometer		
nickel	175–600	2 s and higher
platinum	13–900	
Thermistor	200–570	Less than 0.1 s for flake type
Optical pyrometer	850–3300	Manually adjusted
Infra-red pyrometer	270–3300	Same as thermocouple

12.8 Calibration methods
Primary calibration standards for temperature-measurement devices are achieved relative to the International Practical Temperature Scale

238

(IPTS), at the National Physical Laboratory. Calibration requires the accurate use of specified devices over certain ranges as follows:

13.81 K to 903.89 K – platinum resistance thermometer
903.89 K to 1337.58 K – platinum–platinum/10% rhodium thermocouple
over 1337.58 K – optical pyrometer

These devices are used to give the scales between the fixed points specified in the IPTS.

In industry, calibration and checking is carried out in two ways:

i) fixed-point checking relative to appropriate fixed points on the IPTS;
ii) comparison methods, where the instrument is compared with a secondary standard which has previously been calibrated at the National Physical Laboratory.

12.9 Complete measurement systems

12.9.1 Measurement of exhaust-gas temperatures in an i.c. engine
Figure 12.14 shows a typical arrangement for monitoring exhaust-gas temperatures using a thermocouple. A d.c. amplifier is used to amplify the thermocouple signal to a suitable level for connection to a u.v. galvanometer matching network and recorder. The thermocouple would be enclosed in a protective sheath to prevent corrosion.

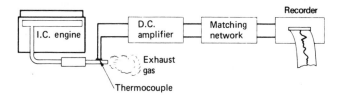

Fig. 12.14 Measurement of exhaust-gas temperature

12.9.2 Measurement of liquid temperature in a heated tank
Figure 12.15 illustrates the use of a platinum resistance thermometer contained in a protective sheath in a tank full of liquid. The resistance change is converted to a direct voltage by means of a d.c.-excited resistance bridge. D.C. amplification provides a voltage magnitude sufficient to drive the pen recorder directly. Note the use of a stirrer to mix the liquid to try to obtain a uniform temperature.

12.9.3 Data-logging system for temperature measurement
The inputs from several thermocouples may be displayed and recorded sequentially by using the data-logging system shown in fig. 12.16. The

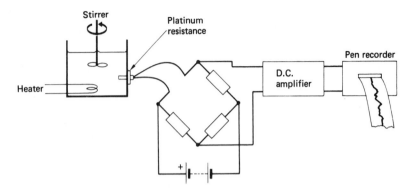

Fig. 12.15 Measurement of temperature in a heated tank

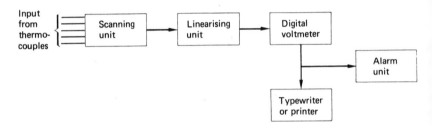

Fig. 12.16 A data-logging system for temperature measurement

scanning unit scans each input in turn, and the lineariser compensates for the non-linearity in the thermocouple characteristic before applying the resulting output voltage to a digital voltmeter (d.v.m.). The d.v.m. has three functions:

a) to display, in digital form, the temperatures of the thermocouples;
b) to provide an output to a pre-set alarm unit;
c) to provide an output voltage to drive a typewriter or printer which provides a permanent record of the temperatures.

12.9.4 Telemetry for temperature measurement
It may sometimes be necessary to measure at some remote point the temperatures within a moving vehicle. One possible solution is shown in fig. 12.17. The voltage from the thermocouple is used to frequency modulate a very-high-frequency radio wave. This radio wave may then be transmitted and received by a radio-frequency receiver which uses a demodulator to extract the original temperature-dependent voltage from the modulated waveform.

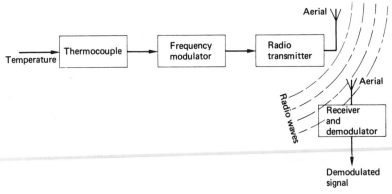

Fig. 12.17 Telemetry for temperature measurement

Exercises on chapter 12

1 What is meant by, and why is it necessary to have, an International Practical Temperature Scale?

2 State, with brief reasons, the method you would choose to measure the temperature of (a) a bath of water, (b) a furnace melting iron, and (c) different parts of a diesel engine.

3 Explain the principle of operation of the platinum resistance thermometer. What are its advantages and disadvantages over the liquid-in-glass thermometer?

4 What is meant by the thermal lag of a thermometer? Why might it be less for a thermocouple than for a resistance thermometer?

5 A type-K thermocouple is exposed to a temperature of 1200°C. If the indicator is used as the cold junction and its temperature is 50°C, use fig. 12.5 to calculate the e.m.f. indicated. [47 mV]

6 At any particular instant, a resistance thermometer indicates a temperature θ_2 of 50°C while the actual temperature θ_1 is 100°C. If the dynamic relationship for the resistance thermometer is given by

$$\frac{d\theta_2}{dt} = k(\theta_1 - \theta_2) \quad \text{where } k = 0.2 \, \text{s}^{-1}$$

determine the time constant for the thermometer. [5 s]

7 Temperature measurement can be divided broadly into two categories: (a) non-electrical methods and (b) electrical methods. List the types of thermometer in each category.

Select one of the electrical methods and describe how temperature measurement may be achieved. In your description include

i) a typical circuit arrangement,
ii) a typical characteristic of the device
iii) the normal temperature range of the device, and
iv) any errors that may arise, and how they may be reduced.

241

13 Introduction to control

13.1 General introduction
All modern engineering systems include certain aspects of control systems at some point. In their broadest sense, control engineering and the associated theory are concerned with the means by which systems may be made to behave in a desired way.

Until recently control theory had only been applied to engineering systems; however, in recent years attempts have been made to extend the theory to such wide-ranging areas as budgetary control, control systems in the human body, management control, and even to the economies of countries and to world dynamics (the interrelationship between populations and resources).

In the field of engineering, control has wide-ranging applications from the simple thermostat arrangement to control the temperature of an oven to the complex computer-controlled sampling arm and soil-analysing equipment used on the Viking lander which was landed on the planet Mars.

13.2 Block-diagram representation of systems
In control engineering, as in the measurements discussed earlier, a diagrammatic method of representing systems is used, known as *block-diagram representation.*

A rectangular block is used to represent the element, and arrows show the direction of signal flow. A flow-measuring system would be represented as shown in fig. 13.1.

Fig. 13.1 Block diagram of a flow-measuring system

In control systems, an additional diagram is required to represent the comparison element that is used in closed-loop systems. A comparison element is represented as shown in fig. 13.2.

We will use this method to represent the relative positions of elements within the control loops. In a mathematical analysis of the system, the

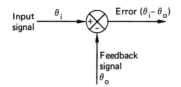

Fig. 13.2 Block diagram of a comparison element

equation (transfer operator) representing the relationship between the output and the input of a particular element would be shown in the block diagram.

Example 13.1   Draw a block diagram to represent the level control system shown in fig. 13.3.

The block diagram will be as shown in fig. 13.4.

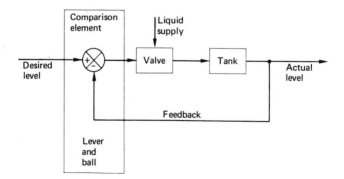

Fig. 13.3 A level-control system

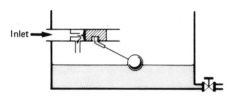

Fig. 13.4 Block diagram of a level-control system

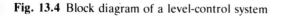

13.3 Types of control system

13.3.1 Open-loop systems (unmonitored control system)

a) Continuous systems

A typical example of a continuous open-loop control system is an electric storage radiator without thermostatic control, as illustrated in fig. 13.5.

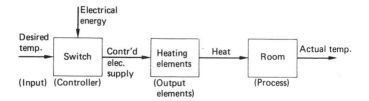

Fig. 13.5 Block diagram of an electric storage radiator

Consider a typical British winter sequence for this type of system. It is a cold night, so the radiator is set at a high value. During the night the weather changes and becomes mild (this external disturbance represents a load change) so the next morning the room is far too hot, consequently the radiator setting is reduced. The following night the weather goes cold again, with the result that the next morning the room is too cold.

In this type of system the main features are

i) there is no comparison between actual and desired values;
ii) each input setting determines a fixed operating position for the controller;
iii) changes in external conditions (load changes) result in an output change unless the controller setting is altered manually.

Since there is no comparison between actual output and the desired value, rapid changes can occur in the output if any load changes occur.

Example 13.2 Draw a block diagram to represent a steam turbine–generator set without a governor.

This would have a block diagram as illustrated in fig. 13.6.

If the system is operating to supply electricity for the national grid, the speed must remain constant to maintain the correct mains frequency. A

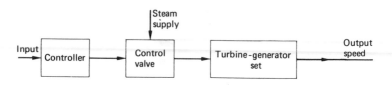

Fig. 13.6 Block diagram of a steam turbine–generator set

call for extra electricity from the generator set causes extra load on the turbine, causing it to slow down, and, unless the input setting is manually changed, the turbine will settle down at a new steady speed.

A typical graph of speed against time, or the system-response curve, would be as shown in fig. 13.7.

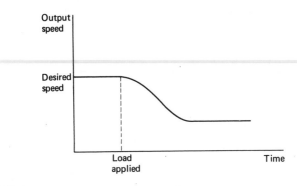

Fig. 13.7 A system-response curve

b) Sequential systems

A sequential-control system is a special type of open-loop system and is employed in many mechanical handling and packaging systems. A typical system would be as shown in fig. 13.8.

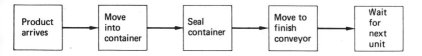

Fig. 13.8 Block diagram of a sequential-control system

In this type of system the main features are

i) the finish of one action initiates the start of the next;
ii) actions take place in a certain fixed sequence;
iii) no comparison of desired and actual values occurs, although feed-back of a completed action is used.

This type of system often employs pneumatic or hydraulic actuation, using cylinders to perform the operations and mechanical trip valves to signal the completion of the operation.

Example 13.3 The drilling rig illustrated in fig. 13.9 is required to perform the following sequence of operations. Pressing a start button

245

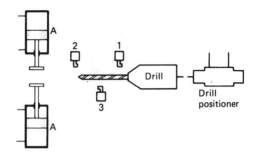

Fig. 13.9 A drilling-rig arrangement

supplies air to the cylinders A to clamp the component, and this initiates the drill movement. The first movement is rapid feed to trip valve 1, followed by slow controlled feed on the drilling stroke to trip valve 2 which causes the drill stroke to reverse. Striking trip valve 3 on the reverse stroke causes cylinders A to unclamp, allowing the workpiece to be removed. Draw a block diagram to represent this sequence of operations.

A block diagram of the system is shown in fig. 13.10.

Fig. 13.10 Block diagram of drilling rig

13.3.2 Closed-loop systems (monitored control systems)

a) On–off systems

Consider the room-heating example in section 13.3.1. To overcome the problem of variations in room temperature due to alterations in external conditions, a thermostat can be fitted in the room. The thermostat then compares the actual room temperature with the desired value and any deviation (error) causes appropriate control action to be taken. In this case the power would be switched off if the temperature was too high and switched on if it was too low. This is known as two-position or on–off control and is illustrated in the block-diagram shown in fig. 13.11.

This type of control is suitable for systems where the changes in load occur slowly and the process is very slow acting, necessitating infrequent changes of control action. When these conditions do not apply, a more suitable type of system is a continuous-control system.

246

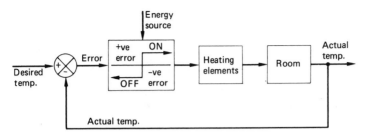

Fig. 13.11 Block diagram of an on–off system

b) Continuous control

In continuous control, as in on–off systems, the actual output of the system is fed back and compared with the desired value in a comparison element which generates a deviation or error signal. In continuous-control systems, however, the controlling signal generated is proportional to the magnitude of the error, hence the term *proportional-control system* is also used.

A typical block-diagram arrangement is shown in fig. 13.12. Note that the energy supply has been omitted from the control element. This is the usual practice for block diagrams.

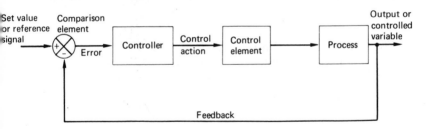

Fig. 13.12 Block diagram of a proportional-control system

Some important features of closed-loop control systems are as follows.

i) Since the actuating signal from the controller only controls the output power, and does not supply it, high-power outputs can be accurately controlled from low-power inputs.

ii) The error signals generated by the comparison element are usually very small in magnitude, hence some form of amplification must be included in the controller for most systems.

iii) Closed-loop systems have a self-regulating property, since any external disturbance will change the output, resulting in an error signal being generated which produces a reaction to maintain the output.

247

The two essentials for a continuous closed-loop system are *feedback* and *comparison*.

Example 13.4   Draw a block diagram to represent a steam turbine-generator set fitted with a speed governor.

This would have a block diagram as illustrated in fig. 13.13.

If the system is again operating to supply electricity for the national grid, a call for extra electricity from the generator set increases the load on the turbine, causing it to slow down. This time, however, the reduced speed is sensed by the governor which compares the actual speed with the desired speed (reference signal) and generates an error or deviation signal. The controller generates a signal so that the control valve is moved an amount proportional to the error. The resulting increased steam supply to the turbine causes it to speed up until the output speed matches the desired speed.

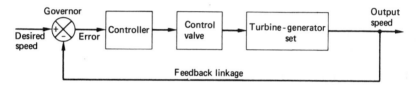

Fig. 13.13 Block diagram of a steam turbine–generator set with governor

The system-response curve would be as shown in fig. 13.14. Note that oscillations occur, because of the system inertia.

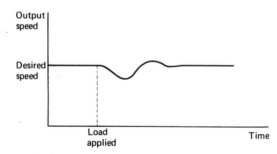

Fig. 13.14 A system-response curve

13.4 Branches of control

13.4.1 Servo-mechanisms (kinetic control)
A servo-mechanism is any control system used for the control of motion parameters such as displacement, velocity, and acceleration. The

248

objective of the control system is to displace the process in such a manner that it follows a continually changing input or desired value (sometimes known as the *demand signal*).

These systems are inherently fast-acting, having very small time lags and response times in the order of milliseconds. Because of the fast response speeds required, this type of system usually employs electric or hydraulic actuation.

Example 13.5 Draw a block diagram to represent the remote position-control system illustrated in fig. 13.15.

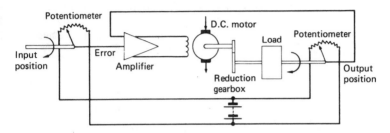

Fig. 13.15 A remote position-control system

This has the block diagram shown in fig. 13.16. Note the typical closed-loop features of feedback, comparison, and error amplification.

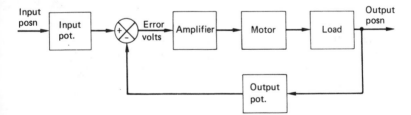

Fig. 13.16 Block diagram of a remote position-control system

Further examples of servo-mechanisms include positioners for radar scanners, machine-tool controls, and automatic pilots.

13.4.2 Process control
Process control is the control of such parameters as pressure, flow, level, temperature, and pH (acidity). In this type of system there is usually only one optimum desired value, known as the *set point*. The control system is required to ensure that the process output is maintained at this level in

spite of changes in external conditions which affect the process (these are given the general name 'load disturbances'). An example of a load disturbance could be a change in boiler steam pressure, affecting a temperature-control system; or a change in raw materials would act as a load disturbance on a mixing process.

These systems are usually slow-acting, with large time lags in the measuring system and the process which may be minutes or even hours. Pneumatic actuation is often employed in this type of system.

Example 13.6 Draw a block diagram to represent the flow-control system illustrated in fig. 13.17.

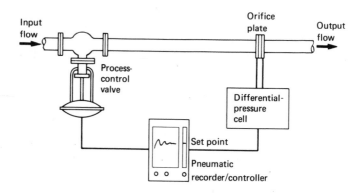

Fig. 13.17 A flow-control system

This has the block diagram shown in fig. 13.18.

Further examples of process-control applications include petro-chemical-plant control, vat-temperature control in breweries, and lime-kiln control.

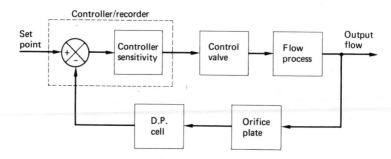

Fig. 13.18 Block diagram of a flow-control system

250

13.4.3 Regulators

Some control systems are classified as regulators, but these are basically particular examples of process control and servo-mechanisms and are concerned with the maintenance of constant controlled variables such as speed, voltage, temperature, etc., irrespective of the external load on the plant.

13.5 Basic control-system considerations

13.5.1 Order of a control system

Automatic-control systems can consist of any number of components of a varied nature – such as electrical, hydraulic, pneumatic, thermal, or mechanical – all operating together in a manner determined by the function of the system. However, all systems, whether process control or servo-mechanisms, can be grouped according to the *order* of the highest differential in the denominator of the system transfer operator, in the same manner as outlined in chapter 2. For example,

$$\frac{\theta_o}{\theta_i} = \frac{5}{1 + 6D} \quad \text{is first-order}$$

$$\frac{\theta_o}{\theta_i} = \frac{10}{D^2 + 3D + 5} \quad \text{is second-order}$$

$$\frac{\theta_o}{\theta_i} = \frac{20}{(1 + 0.1D)(1 + 0.2D)} \quad \text{is also second-order}$$

13.5.2 General feedback systems

A control system can be represented as shown in fig. 13.19. Using G to represent feedforward elements and H to represent feedback elements, we have fig. 13.20.

From fig. 13.20,

$$\epsilon = \theta_i - \gamma_o$$

$$\theta_o = G\epsilon$$

$$\therefore \quad \theta_o = G(\theta_i - \gamma_o)$$

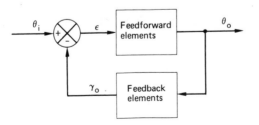

Fig. 13.19 A general control system (θ_i = reference signal or input, θ_o = output signal, ϵ = error signal, γ_o = conditioned output)

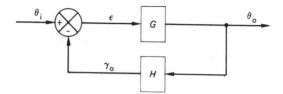

Fig. 13.20 Block diagram of a typical control system

But $\quad \gamma_0 = H\theta_0$

$\therefore \quad\quad \theta_0 = G\theta_i - GH\theta_0$

Rearranging,

$$(1 + GH)\theta_0 = G\theta_i$$

$\therefore$ closed-loop transfer operator $= \dfrac{\theta_0}{\theta_i} = \dfrac{G}{1 + GH}$ 13.1

If θ_i and θ_0 are physically the same – i.e. both displacements or both voltages proportional to position – then the feedback element H is not needed and the feedback is known as *unity feedback*.

Therefore, for unity feedback, H is 1 in equation 13.1 and the closed-loop transfer operator is

$$\frac{\theta_0}{\theta_i} = \frac{G}{1 + G}$$

For an open-loop system we have

$$\frac{\theta_0}{\theta_i} = G \quad\quad\quad\quad 13.2$$

Comparing equations 13.1 and 13.2, it can be seen that the effect of negative feedback is to reduce the overall system gain from G to $G/(1 + GH)$.

Example 13.7 The forward-path gain of an open-loop system is 50. Determine the gain if negative feedback is employed with a feedback element of gain 10.

$$\text{Closed-loop gain} = \frac{G}{1 + GH}$$

$$= \frac{50}{1 + (50 \times 10)}$$

$$\simeq \frac{50}{500} = 0.1$$

13.5.3 Effect of lags on the control system

The elements in a control system cannot respond immediately to input signals; for example, an electric motor has inertia which opposes changes of motion. The element output will, therefore, lag behind the input. Consider a sine wave being injected into a control system like that shown in fig. 13.20. The output of the feedback element γ_o will generally lag behind θ_i as illustrated in fig. 13.21.

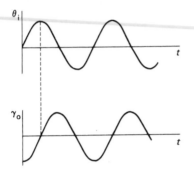

Fig. 13.21 Lag between output and input

It is possible, depending on the number and nature of the elements around the control loop, for the frequency to be such that the feedback signal γ_o lags behind the input signal θ_i by 180°. Now 180° phase lag is equivalent to a gain of -1, therefore instead of γ_o being subtracted from θ_i in the comparison element it will be added to θ_i:

$$\theta_i - (-\gamma_o) = \theta_i + \gamma_o$$

This condition is referred to as *positive feedback*

The magnitude of γ_o is important:

a) If γ_o is less than ϵ, some of the θ_i signal is still required to maintain the output θ_o and the system is therefore still under control; i.e. it is *stable*.

b) If $\gamma_o = \epsilon$, a constant-amplitude signal will pass round the loop and output θ_o will therefore be a constant-amplitude sinusoidal signal; i.e. the system is *marginally stable*.

c) If γ_o is greater than ϵ, the signal will gradually build up as it passes round the loop; i.e. the system is *unstable*.

The positive-feedback condition can be represented as shown in fig. 13.22.

From fig. 13.22,

$$\epsilon = \theta_i + \gamma_o$$

$$\theta_o = G\epsilon$$

$$\therefore \quad \theta_o = G\theta_i + G\gamma_o$$

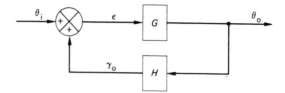

Fig. 13.22 Block diagram for positive feedback

But $\gamma_o = H\theta_o$

$\therefore \quad \theta_o = G\theta_i + GH\theta_o$

Rearranging,

$$(1 - GH)\theta_o = G\theta_i$$

$\therefore \quad$ closed-loop transfer operator $= \dfrac{\theta_o}{\theta_i} = \dfrac{G}{1 - GH}$

Exercises on chapter 13

1 Explain briefly the meaning of the following terms when used with reference to a closed-loop control system: (a) kinetic control, (b) process control, (c) regulators, (d) transfer operator, and (e) order of the system.

2 Sketch a block diagram of a typical closed-loop automatic-control system and indicate the necessary three basic elements of such a system. By describing the progression of a signal around the control loop, explain in detail how the control function is achieved and summarise the essential features of the system.

3 Differentiate between the terms 'open' and 'closed' loop when applied to a control system. Illustrate your answer by reference to a particular example of each type of system and sketch its relevant block diagram.

4 Sketch a block diagram to represent a thermostatic temperature-control system for an oven and state what type of system it is.

5 An automatic radar tracker contains a motor driving an aerial through a reduction gearbox. The motor is driven by an amplifier, the signal to which is a voltage proportional to the error between the desired and actual aerial positions. Sketch a block diagram to represent the system.

6 A control system contains an input transducer, output transducer, comparator, error amplifier, and actuator. Draw a labelled block diagram showing these components arranged for the control of a load, and describe the function of each component.

7 The control system representing a man driving a car can be described as follows. The direction or heading of the road may be considered as the input; the actual direction of the car is the output of the system. The driver controls the output by constantly measuring it with his eyes and brain and correcting the direction of the car by means of the steering wheel. Sketch a block diagram to represent the system.

8 The direction-control system for a guided missile operates as follows. The relative directions of the missile and target are measured using a gyroscope in the missile, and the error is fed to a controller which operates a servo-motor to deflect the rocket jet, thus altering the missile's path. Sketch a block diagram for the system.

9 A bin-tipper mechanism is activated when a bin arrives on a conveyor and actuates a trip valve. The bin is emptied, returned to the upright position, and moved on down the conveyor to be refilled and returned to the tipper. Sketch a block diagram representing this system and state what type of system it is.

10 The forward-path gain of an open-loop system is 100. If the system is made into a closed-loop system with negative feedback gain of 2.5, determine the percentage reduction in the overall gain. [99.6%]

14 Control-system components

Automatic-control systems, including their recording elements, may be represented by a general block diagram as shown in fig. 14.1.

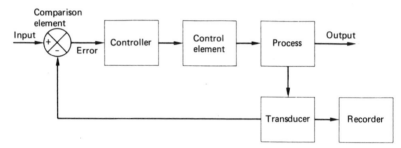

Fig. 14.1 A general control system

Transducers and recorders have been dealt with in earlier chapters, and most types of controllers are beyond the scope of this book – as far as we are concerned, the controller is considered as a device producing an amplified and conditioned signal proportional to the error. The controller can therefore be represented as a block diagram as shown in fig. 14.2.

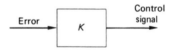

Fig. 14.2 Block-diagram representation of a controller (K is a constant)

Some common examples of comparison elements and control elements are considered in the following sections.

14.1 Comparison elements
Comparison elements compare the output or controlled variable with the desired input or reference signal and generate an error or deviation signal. They perform the mathematical operation of subtraction.

14.1.1 *Differential levers (walking beams)*

Differential levers are mechanical comparison elements which are used in many pneumatic elements and also in hydraulic control systems. They come in many varied and complex forms, a typical example being illustrated in fig. 14.3, which shows a type used in a Taylor's Transcope pneumatic controller.

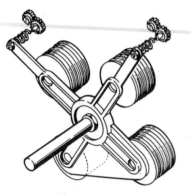

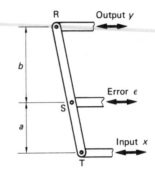

Fig. 14.3 The motion plate for a Taylor Transcope controller

Fig. 14.4 The differential lever

For purposes of analysis a differential lever can be considered as a simple lever which is free to pivot at points R, S, and T as illustrated in fig. 14.4.

From fig. 14.4 for *small* movements,

i) considering R fixed: if x moves to the right then

$$\epsilon = \frac{b}{a+b} x$$

ii) considering T fixed: if y moves to the left then

$$\epsilon = - \frac{a}{(a+b)} y \quad \text{(movement to the right taken as +ve)}$$

The total movement ϵ can be found by using the principle of superposition, which states that, for a linear system, the total effect of several disturbances can be obtained by summation of the effects of each individual disturbance acting alone. The total movement ϵ due to the motion of x and y is therefore given by the sum of (i) and (ii):

$$\epsilon = \frac{b}{(a+b)} x - \frac{a}{(a+b)} y$$

In many cases it is arranged that $a = b$, so that the lever is symmetrical, and then

$$\epsilon = \tfrac{1}{2}(x - y)$$

i.e. $\epsilon = \tfrac{1}{2} \times$ error   or $\tfrac{1}{2} \times$ deviation

It is important that the output movement at y is arranged to always be in the opposite direction to the input x, i.e. a negative-feedback arrangement.

Example 14.1   Calculate the value of ϵ for the differential lever in fig 14.4, if $a = 2b$.

$$\epsilon = \frac{b}{2b + b}x - \frac{2b}{(2b + b)}y$$

$$= \tfrac{1}{3}(x - 2y)$$

Thus it is only with $a = b$ that a true error signal is obtained.

14.1.2 Potentiometers
Potentiometers are used in many d.c. electrical positioning servo-systems. They consist of a pair of matched resistance potentiometers operating on the null-balance principle. The sliders are driven by the input and output shafts of the control system as illustrated in fig. 14.5.

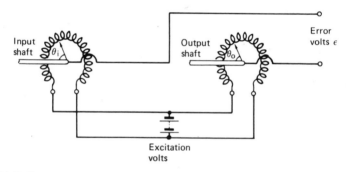

Fig. 14.5 Error detection by potentiometers

If the same voltage is applied to each of the potentiometer windings, an error voltage is generated which is proportional to the relative positions of the wipers. We have

$$\epsilon = K_p(\theta_i - \theta_o)$$

where θ_i = input-shaft position

   θ_o = output-shaft position

and K_p = potentiometer sensitivity (volts/degree)

When the input and output shafts are aligned, $\theta_i = \theta_o$ and the error voltage ϵ is zero; i.e. null balance is achieved.

Some of the limitations of wire-wound potentiometers have already been discussed in chapter 3 – in particular, the noise generated as the slider moves across the wire can be a source of inaccuracy when they are used in control systems. To minimise this, potentiometers with very high resolution must be employed.

A further problem encountered when simple single-turn potentiometers are used is that the maximum slider sweep must be limited to about 330°, otherwise dead-zones (areas of no signal) will be present. Servo-systems requiring continuous rotations greater than 360° need specially designed potentiometers, such as single-turn sawtooth potentiometers employing internal electronic switching to overcome the dead-zone. Another method is to use special multi-tapped coils.

Example 14.2 A precision conductive-plastics potentiometer has the following items quoted in its specification:

a) actual electrical travel $355° \pm 2°$
b) resistance $2\,k\Omega$
c) resolution 0.002%
d) power rating   2.5 W at $+40°C$, derating to 0 W at $+110°C$
e) maximum applied volts at 2.5 W load   70 V

Two of these potentiometers are used on the input and output shafts of a position servo-system. If a voltage equal to 75% of the maximum allowable is applied to the coils of both potentiometers, determine (i) the potentiometer sensitivity and (ii) the expected error voltage if there is a misalignment of 3° between the input and output shaft positions.

i)   Voltage applied $= 0.75 \times 70\,V$

$$= 52.5\,V$$

Potentiometer sensitivity $K_p = \dfrac{\text{applied volts}}{\text{electrical travel}}$

$$= \frac{52.5\,V}{355\ \text{degrees}}$$

$$= 0.15\,V/\text{degree}$$

ii)   Error voltage $= K_p \times \text{misalignment}$

$$= 0.15\,V/\text{degree} \times 3°$$

$$= 0.45\,V$$

14.1.3 Synchros

Synchros are the a.c. equivalent of potentiometers and are used in many a.c. electrical systems for data transmission and torque transmission for driving dials. They are also used to compare input and output rotations in a.c. electrical servo-systems and rotating hydraulic systems.

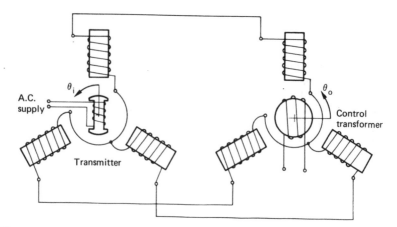

Fig. 14.6 Error detection by synchros

To perform error detection, two synchros are used: one in the mode of a control transmitter and the other as a control transformer, as shown in fig. 14.6.

The synchros have their stator coils equally spaced at 120° intervals. An a.c. voltage (often 115 V at 400 Hz) is applied to the transmitter rotor, producing voltages in the stator coils (by transformer action) which uniquely define the angular position of the rotor. These voltages are transmitted to the stator coils of the transformer, producing a resultant magnetic field aligned in the same direction as the transmitter rotor.

The transformer rotor acts as a 'search coil' in detecting the direction of its stator field. The maximum voltage is induced in the transformer rotor coil when the rotor axis is aligned with the field. Zero voltage is induced when the rotor axis is perpendicular to the field. The 'in-line' position of the input and output shafts therefore requires the transformer rotor coil to be at 90° to the transmitter rotor coil.

The output voltage is an amplitude-modulated signal which requires demodulating to produce the following relationship for small mis-alignment angles:

$$\text{output voltage} = K(\text{input-shaft position} - \text{output-shaft position})$$
$$= K(\theta_i - \theta_o)$$

where K = voltage gradient (volts/degree)

Compared to d.c. potentiometers, synchros have the following advantages:

a) a full 360° of shaft rotation is always available;
b) since they have no sliding contacts, their life expectancy is much higher, resolution is infinite, and hence they do not have 'noise' problems;

c) a.c. amplifiers can be employed and therefore there are no drift problems

However, phase-sensitive rectifiers are necessary to sense direction.

14.1.4 Operational amplifiers

Operational amplifiers, or 'op. amps', are direct-coupled (d.c.) amplifiers with special characteristics as outlined in chapter 4. They can be used for error detection by employing the circuit arrangement shown in fig. 14.7.

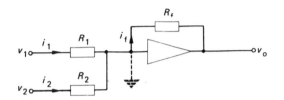

Fig. 14.7 Error detection by an operational amplifier

Applying the same analysis as in chapter 4, the input current to the amplifier can be assumed to be negligible, and

$$i_1 + i_2 = i_f$$

$$\therefore \quad \frac{v_1 - 0}{R_1} + \frac{v_2 - 0}{R_2} = \frac{0 - v_o}{R_f}$$

and
$$v_o = - \left[\frac{R_f}{R_1} v_1 + \frac{R_f}{R_2} v_2 \right]$$

If $R_f = R_1 = R_2$, v_1 is made equal to input (θ_i), and v_2 is made equal to $-$ output ($-\theta_o$), we have

$$v_o = -(\theta_i - \theta_o)$$
$$= -(\text{error})$$

The negative sign can be removed by using an inverter, as shown in example 14.3.

Operational amplifiers are used in electrical control systems and as comparison elements in many hydraulic positioning systems.

Example 14.3 In fig. 14.7, $R_f = 1\,\text{M}\Omega$, $R_1 = R_2 = 0.1\,\text{M}\Omega$, v_1 is a voltage proportional to the input displacement θ_i, and v_2 is a voltage proportional to the output displacement θ_o and is arranged to be fed back in a negative sense. Assuming the proportional constant is 1 V/degree, determine the amplification through the op. amp and show how the sign of the error output can be inverted.

261

We have $$v_o = -\left[\frac{1\,\mathrm{M}\Omega}{0.1\,\mathrm{M}\Omega}\,\theta_i - \frac{1\,\mathrm{M}\Omega}{0.1\,\mathrm{M}\Omega}\,\theta_o\right]$$

$$= -10(\theta_i - \theta_o)$$

The amplification is therefore 10.

The sign of the error can be inverted as shown in fig. 14.8.

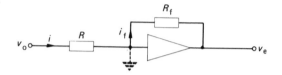

Fig. 14.8 An inverter

We have $i = i_f$

$$\therefore \quad \frac{v_o - 0}{R} = \frac{0 - v_e}{R_f}$$

$$\therefore \quad v_e = -\frac{R_f}{R}v_o$$

and, if R_f is made equal to R,

$$v_e = -v_o$$

14.2 Control elements

Control elements are those elements in which the amplified and conditioned error signal is used to regulate some energy source to the process.

14.2.1 Process-control valves

In many process systems, the control element is the pneumatically actuated control valve, illustrated in fig. 14.9, which is used to regulate the flow of some fluid.

A control valve is essentially a pressure-reducing valve and consists of two major parts: the valve-body assembly and the valve actuator.

a) Valve actuators

The most common type of valve actuator is the pneumatically operated spring-and-diaphragm actuator illustrated in fig. 14.9, which uses air pressure in the range 0.2 bar to 1 bar unless a positioner is used which employs higher pressure to give larger thrusts and quicker action. The air can be applied to the top (air-to-close) or the bottom (air-to-open) of the diaphragm, depending on the safety requirements in the event of an air-supply failure.

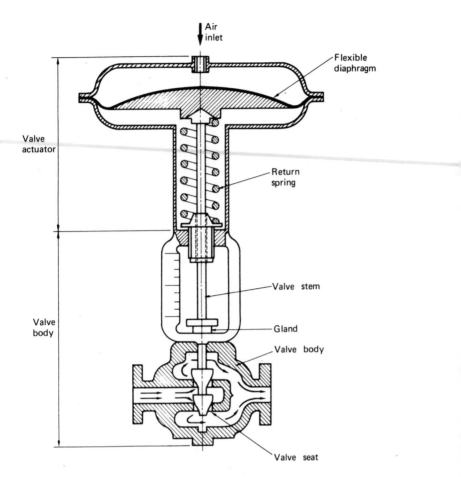

Fig. 14.9 A process-control valve

b) Valve-body design

Most control-valve bodies fall into two categories: single-seated and double-seated.

i) Single-seated valves have a single valve plug and seat and hence can be readily designed for tight shut-off with virtually zero flow in the closed position. Unless some balancing arrangement is included in the valve design, a substantial axial stem force can be produced by the flowing fluid stream.

ii) Double-seated valves have two valve plugs and seats, as illustrated in fig. 14.9. Due to the fluid entering the centre and dividing in both upward and downward directions, the hydrodynamic effects of fluid pressure tend to cancel out and the valves are said to be 'balanced'.

263

Due to the two valve openings, flow capacities up to 30% greater than for the same nominal size single-seat valve can be achieved. They are, however, more difficult to design to achieve tight shut-off.

The valve plugs and seats – known as the valve 'trim' – are usually sold as matched sets which have been ground to a precise fit in the fully closed position.

The valve plugs are of two main types: the solid plug and the skirted V-port plug, as illustrated in fig. 14.10. All valves have a throttling action which causes a reduction in pressure. If the pressure increases again too rapidly, air bubbles entrained in the fluid 'implode', causing rapid wear on the valve plugs. This process is known as *cavitation*. The skirted V-port plugs have less tendency to cause this rapid pressure recovery and are therefore less prone to cavitation.

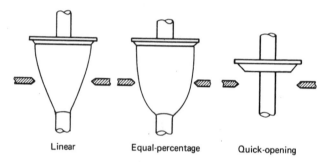

Linear   Equal-percentage   Quick-opening

(a)   Solid   plugs

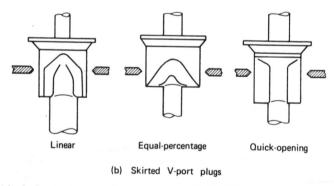

Linear   Equal-percentage   Quick-opening

(b)   Skirted   V-port   plugs

Fig. 14.10 Control-valve plugs

c) Valve flow characteristics
The flow characteristic of a valve is the relationship between the rate of flow change and the valve lift. The characteristics quoted by the manu-

facturers are theoretical or inherent flow characteristics obtained for a constant pressure drop across the valve. The actual or installed characteristics are different from the inherent characteristics since they incorporate the effects of line losses acting in series with the pressure drop across the valve. The larger the line losses due to pipe friction etc., the greater the effect on the characteristic.

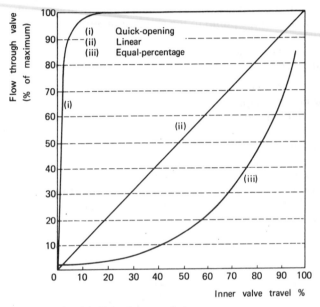

Fig. 14.11 Types of valve flow characteristic

The three main types of characteristic illustrated in fig. 14.11 are:

i) Quick-opening – the open port area increases rapidly with valve lift and the maximum flow rate is obtained after about 20% of the valve lift. This is used for on–off applications.

ii) Linear – the flow is directly proportional to valve lift. This is used when the pressure drop across the valve is reasonably constant, for example in bypass service of pumps and compressors.

iii) Equal-percentage – the change in flow is proportional to the rate of flow just before the flow change occurred; that is, an equal percentage of flow change occurs per unit valve lift. This is used when major changes in pressure occur across the valve and where there is limited data regarding flow conditions in the system.

Example 14.4 A pneumatically controlled process valve is to be used to control the flow of cooling water to a heat exchanger. Explain what type of actuator is required for fail-safe considerations.

For a cooling system, if the air to the valve fails, the valve must open and allow full flow to the heat exchanger rather than close and thus remove all cooling from the system. An air-to-close actuator would therefore be required.

14.2.2 Hydraulic servo-valves

In hydraulic control systems, the hydraulic energy from the pump is converted to mechanical energy by means of a hydraulic actuator. The flow of fluid from the pump to the actuator in most systems is controlled by a servo-valve.

A servo-valve is a device using mechanical motion to control fluid flow. There are three main modes of control:

   i) sliding – the spool valve;
  ii) seating – the flapper valve;
 iii) flow-dividing – the jet-pipe valve.

a) Spool valves

Spool valves are the most widely used type of valve. They incorporate a sliding spool moving in a ported sleeve as illustrated in fig. 14.12. The valves are designed so that the output flow from the valve, at a fixed pressure drop, is proportional to the spool displacement from the null position.

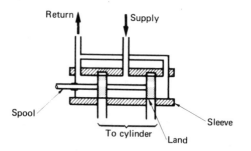

Fig. 14.12 A spool valve

Spool valves are classified according to the following criteria.

The *number of 'ways'* flow can enter or leave the valve. A four-way valve is required for use with double-acting cylinders.

The *number of lands* on the sliding spool. Three and four lands are the most commonly used as they give a balanced valve, i.e. the spool does not tend to move due to fluid motion through the valve.

The *valve-centre characteristic*, i.e. the relationship between the land width and the port opening. The flow–movement characteristic is directly related to the type of valve centre employed. Figure 14.13 illustrates the characteristics of the three possibilities discussed below.

266

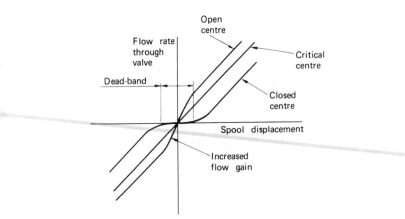

Fig. 14.13 Valve-centre characteristics

   i) Critical-centre or line-on. The land width is exactly the same size as the port opening. This is the ideal characteristic as it gives a linear flow–movement relationship at constant pressure drop. It is very difficult to achieve in practice, however, and a slightly overlapped characteristic is usually employed.

  ii) Closed-centre or overlapped. The land width is larger than the port opening. If the overlap is too large, a dead-band results, i.e. a range of spool movement in the null position which produces no flow. This produces undesirable characteristics and can lead to steady-state errors and instability problems.

 iii) Open-centre or underlapped. The land width is smaller than the port opening. This means that there is continuous flow through the valve, even in the null position, resulting in large power losses. Its main application is in high-temperature environments which require a continuous flow of fluid to maintain reasonable fluid temperatures.

b) Flapper valves
Flapper valves incorporate a flapper–nozzle arrangement as outlined in section 3.15.3. They are used in low-cost single-stage valves for systems requiring accurate control of small flows. A typical arrangement is illustrated in fig. 14.14, which shows a Dowty series 4568 single-stage valve.

    Control of flow and pressure in the service line is achieved by altering the position of the diaphragm relative to the nozzle, by application of an electrical input current to the coil. Increasing the nozzle gap causes a reduction in service-port pressure, since the flow to the return line is increased.

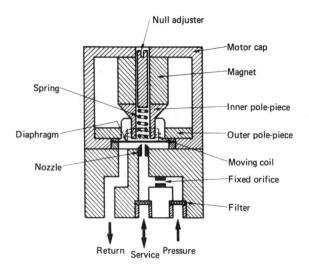

Fig. 14.14 A Dowty single-stage servo-valve

c) Jet-pipe valves
Jet-pipe valves employ a swivelling-jet arrangement and are only used as the first stage of some two-stage electrohydraulic spool valves.

d) Two-stage electrohydraulic servo-valves
These are among the most commonly used valves. A typical arrangement is illustrated in fig. 14.15, which shows a Dowty series 4551 'M' range servo-valve. This incorporates a double flapper–nozzle arrangement as the first stage, driving the second-stage spool.

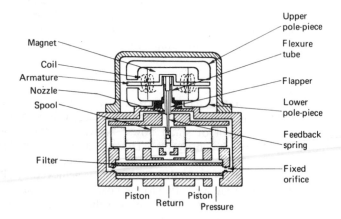

Fig. 14.15 A Dowty electrohydraulic servo-valve

The flapper of the first-stage hydraulic amplifier is rigidly attached to the mid-point of the armature and is deflected by current input to the coil. The flapper passes between two nozzles, forming a double flapper–nozzle arrangement so that, as the flapper is moved, pressure increases at one nozzle while reducing at the other. These two pressures are fed to opposite ends of the main spool, causing it to move.

The second stage is a conventional four-way four-land sliding spool valve. A cantilever feedback spring is fixed to the flapper and engages a slot at the centre of the spool. Spool displacement causes a torque in the feedback wire which opposes the original input-signal torque on the armature. Spool movement continues until these two torques are balanced, when the flapper, with the forces acting on it in equilibrium, is restored to its null position between the nozzles.

e) General considerations

All servo-valves are made to very fine tolerances – of the order of 0.0075 mm – and hence tend to be expensive. These tolerances necessitate very fine filtration, down to a few micrometres particle size, hence all hydraulic systems must incorporate filters, and quite often internal filters are built into the servo-valve.

f) Transfer operators of servo-valves

The transfer operator of most servo-valves can usually be approximated by a second-order lag with natural frequencies in the order of 80 Hz. However, for the simple analysis that we will carry out later, the transfer operator is approximated as a zero-order element.

Example 14.5 A Dowty series 4551 servo valve has the following quoted specifications:

a) flow range   9.6 litres/min
b) maximum supply pressure   315 bars
c) rated input signal   15 mA
d) frequency response at -3 dB amplitude ratio   100 Hz
e) system filtration   10 μm recommended

Explain the meaning of these terms.

a) This is the load flow through the valve for a standard pressure drop through the valve of 70 bars at the rated input current.
b) This is the maximum supply pressure to be used on the valve.
c) This is the differential input current between the first-stage motor coils to produce the rated flow.
d) This is the bandwidth of the valve and represents the frequency over which the valve gives a reasonable response, i.e. from d.c. to 100 Hz.
e) This is the minimum grade of filter required. It must be able to remove particles down to 10×10^{-6} m.

14.2.3 Hydraulic actuators

As stated in the previous section, the hydraulic servo-valve is used to control the flow of high-pressure fluid to hydraulic actuators. The hydraulic actuator converts the fluid pressure into an output force or torque which is used to move some load.

There are two main types of actuator: the rotary and the linear, the latter being the most commonly used.

Linear actuators are commonly known as rams, cylinders, or jacks, depending on their application. For most applications a double-acting cylinder is required – these have a port on each side of the piston so that the piston rod can be powered in each stroke direction, enabling fine control to be achieved. A typical cylinder design is shown in fig. 14.16.

Fig. 14.16 A linear actuator

Example 14.6   Figure 14.17 shows a diagrammatic hydraulic servo-valve/cylinder arrangement. Assuming that the flow through the valve is directly proportional to the valve spool movement, and neglecting leakage and compressibility effects in the cylinder, derive a simple transfer operator for this system.

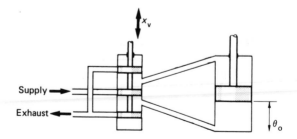

Fig. 14.17 A servo-valve/cylinder arrangement

Referring to fig. 14.17:

For the servo-valve,

$$\text{volumetric flow rate through the valve } \dot{v} \propto \text{valve spool movement } x_v$$

$$\therefore \qquad \dot{v} = K_v x_v$$

where K_v = valve characteristic

For the cylinder,

$$\text{volumetric flow rate to the cylinder } \dot{v} = \text{effective cylinder area} \times \text{piston velocity}$$

$$\therefore \qquad \dot{v} = A \times \frac{d\theta_o}{dt}$$

Using the D operator, we have

$$\dot{v} = AD\theta_o$$

Substituting for $\dot{v}$, we get

$$K_v x_v = AD\theta_o$$

Therefore the transfer operator is

$$\frac{\theta_o}{x_v} = \frac{K_v}{AD} \qquad \text{i.e. an integrator, since } 1/D \equiv \int dt$$

14.2.4 D.C. servo-motors

D.C. servo-motors have the same operating principle as conventional d.c. motors but have special design features such as high torque and low inertia, achieved by using long small-diameter rotors.

Two methods of controlling the motor torque are used:

a) field control – fig. 14.18(a);
b) armature control – fig. 14.18(b).

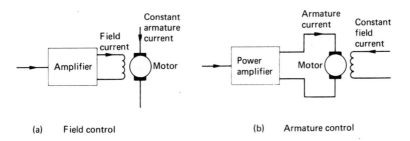

(a)   Field control   (b)   Armature control

Fig. 14.18 Control of d.c. servo-motors

a) Field control

With field control, the armature current is kept approximately constant and the field current is varied by the control signal. Since only small currents are required, this means that the field can be supplied direct from electronic amplifiers; hence the special servo-motors are wound with a split field and are driven by push–pull amplifiers.

Most of these systems are damped artificially by means of velocity feedback, which requires a voltage proportional to speed. This is achieved by means of a tachogenerator which is built with the motor in a common unit.

Field-controlled d.c. motors are used for low-power systems up to about 1.5 kW and have the advantage that the control power is small and the torque produced is directly proportional to the control signal; however, they have a relatively slow speed of response.

b) Armature control

With armature control, the field current is kept constant and the armature current is varied by the control signal.

Considerable development has taken place in the design of this type of motor for use in robot drive systems. A common form in use is the disc armature motor (sometimes called a pancake motor). This consists of a permanent magnet field and a thin disc armature consisting of copper tracks etched or laminated onto a non-metallic surface. These weigh less than conventional iron-core motors giving very good power to weight ratios and hence a fast speed of response.

Power outputs in the range 0.1 to 10 kw are typical.

14.2.5 A.C. servo-motors

A.C. servo-motors are usually two-phase induction motors with the two stator coils placed at right angles to each other as shown schematically in fig. 14.19. The current in one coil is kept constant, while the current in the other coil is regulated by the amplified control signal. This arrangement gives a linear torque/control-signal characteristic over a limited working range.

They are usually very small low-power motors, up to about 0.25 kW.

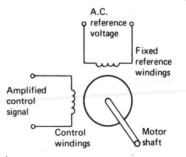

Fig. 14.19 A two-phase a.c. servo-motor

As with the d.c. motors in the previous section, servo-motor/tachogenerator units are supplied to facilitate the application of velocity feedback.

14.3 Analysis of simple control systems
The analysis of some basic control systems into their component parts and the identification of their individual functions is illustrated in the following examples.

14.3.1 Electrohydraulic servo-systems
Figure 14.20 shows the arrangement of a hydraulic system for manually operating an aerodynamic control surface.

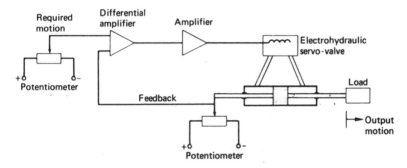

Fig. 14.20 An electrohydraulic servo-system

a) The input and output resistance potentiometers are transducers for converting linear displacement into a voltage.
b) The differential amplifier is the comparison element generating the error signal.
c) The amplifier is the controller producing an amplified error signal.
d) The electrohydraulic servo-valve is the control element, controlling the flow of high-pressure oil to the actuator which moves the load.

14.3.2 A continuous weighing system
Figure 14.21 shows an arrangement of an electrohydraulic system for controlling the movement of material from a hopper along a conveyor belt.

a) The electronic controller/recorder acts as the comparison element and controller, producing a small current signal proportional to the error between the set point (the desired material flow) and the actual flow.
b) This current signal drives the control element (in this case a servo-valve) which, by activating a cylinder, drives a slide to regulate the flow of the material.
c) The flow of material along the conveyor is the process.
d) The electronic weighing machine acts as the output transducer and

273

signal conditioner, producing a signal proportional to material flow suitable for feedback to the controller.

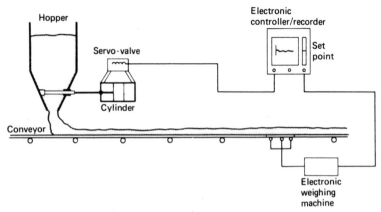

Fig. 14.21 A continuous weighing system

14.3.3 A temperature-control system

Figure 14.22 shows the arrangement of a control system for controlling the temperature in a jacketed kettle by differentially regulating two valves in the water and steam lines.

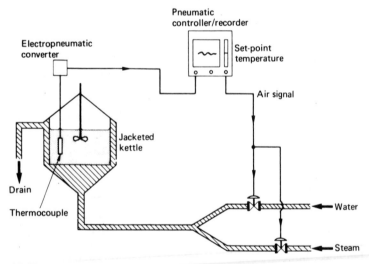

Fig. 14.22 A temperature-control system

a) The pneumatic controller/recorder acts as the comparison element and controller, producing an air signal proportional to error.
b) The pneumatically actuated valves are the control elements.

274

c) The reaction taking place in the jacketed kettle is the process.
d) The thermocouple is the output transducer, producing a small voltage signal proportional to temperature.
e) The electropneumatic converter is the signal conditioner, producing from the voltage signal an air signal in the range 0.2 bar to 1 bar.

14.3.4 A power-steering unit
Figure 14.23 shows an arrangement of a power-steering unit for a car.

a) The steering-wheel rotary motion is the input signal and the rack-and-pinion mechanism acts as the input signal conditioner which converts rotary motion into linear motion.
b) The differential lever is the comparison element which produces a signal proportional to error to move the spool of the servo-valve.
c) The servo-valve is the control element, regulating flow to the power piston.
d) The process is the transmission of force to move the wheels.

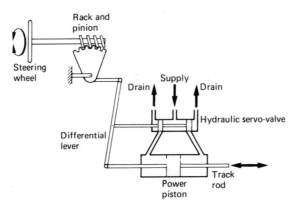

Fig. 14.23 A power-steering unit

14.4 Proposing a control system
A typical procedure for deciding a suitable method of automatic control for a process or system is detailed below.

a) Identify the input and output for the system.
b) Decide on the type of transducer required for measuring the output.
c) Determine the type of signal conditioning necessary for the trans-duced output (and possibly input) so that a suitable comparison element can be arranged.
d) Decide on the type of system necessary, e.g. electrical, hydraulic, pneumatic, or electrohydraulic, etc.
e) Determine the type of control elements required.

Example 14.7 Propose a mechanical–hydraulic control system suitable for a reproducing shaper so that a duplicating cutter follows the position of the master cutter.

a) The input and output are the master and duplicating cutter positions, respectively.
b) In a simple system, direct positions can be used.
c) A mechanical comparison element can be used, i.e. a differential lever.
d) The type of system is specified.
e) A hydraulic spool valve and cylinder will be suitable.

Figure 14.24 shows a simple arrangement for the system. The lever shown is included to reverse the direction of the output position θ_o so that negative feedback is applied to the differential lever.

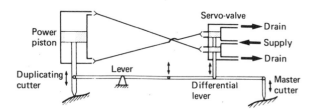

Fig. 14.24 A reproducing shaper

Example 14.8 Propose a control system to maintain the thickness of plate produced by the final stand of rollers in a steel-rolling mill.

a) The input will be the desired plate thickness and the output will be the actual thickness.
b) The required thickness will be set by a dial control incorporating a position transducer which produces an electrical signal proportional to the desired thickness. The output thickness will have to be measured using a device such as a β-ray thickness gauge with amplification to provide a suitable proportional voltage.
c) With two voltage signals, an operational amplifier will be suitable as a comparison element.
d) The desired power for moving the nip roller will require hydraulic actuation.
e) A power piston regulated by an electrohydraulic servo-valve will be suitable.

Figure 14.25 shows a simple arrangement for the system.

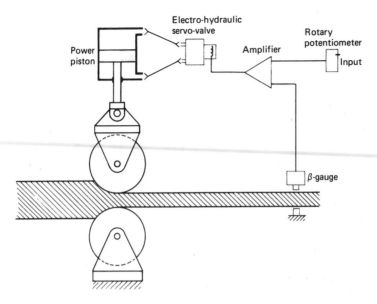

Fig. 14.25 Thickness control

Example 14.9 Propose a control system to maintain a fixed fluid level in a tank. (It must be possible to vary the level setting easily.) The flow is to be regulated on the input side, and the output from the tank is flowing into a process with a variable demand.

a) The input will be the desired fluid level and the output the actual level.
b) Since the output is a variable level, a capacitive transducer will be suitable.
c) Since the system is a process-type system, a commercial controller will be suitable and the desired level will therefore be a set-point position on the controller. If a pneumatic controller is chosen, the electrical signal from the capacitive level transducer will have to be converted into a pneumatic signal by means of an electropneumatic converter.
d) The choice of a pneumatic controller means that the system will be electropneumatic.
e) A suitable control element will be an air-to-open pneumatically actuated control valve.

Figure 14.26 shows a simple arrangement for the system.

Exercises on chapter 14
1 Suggest ways in which the following parameters could be compared for control purposes: (a) two displacements, (b) two velocities, (c) two voltages, (d) two angular velocities.

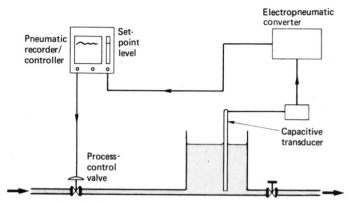

Fig. 14.26 Level control

2 Sketch a circuit using an operational amplifier which would generate a controlling signal equal to 100 times the error signal between an input and output voltage of a servo-system.

3 Explain the difference between inherent and installed characteristics when discussing process-control valve flow characteristics.

4 A pneumatically controlled process valve is to be used to control the flow of oil to a boiler furnace. Explain what type of actuator is required for fail-safe considerations.

5 An electrohydraulic servo-valve is described in a catalogue as '. . . a four-way, four-land, closed-centre valve'. Explain, with the aid of sketches, the meaning of these terms.

6 Draw block diagrams representing the systems described in sections 14.3.1 to 14.3.4.

7 Figure 14.27 shows an arrangement of an industrial heating and cooling system. Analyse the system into its component parts and identify the function of each.

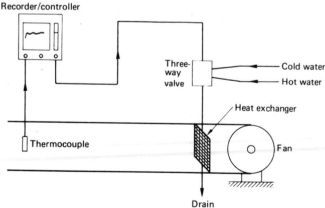

Fig. 14.27 An air-conditioning system

15   System responses

15.1 Introduction

System responses have been discussed in chapter 2 with reference to system performance. In this chapter they will be dealt with qualitatively to a greater depth, the detailed mathematics being covered in the appendices.

   System responses consist of two parts:

i) the transient response, which – if the system is stable – will die away to give

ii) the steady-state response.

   The input to any measuring or control system will in general be of a random nature; however, the three standard inputs mentioned in chapter 2 will be considered. These are:

a) Step or constant-value input. This provides information about the transient response of the system.

b) Ramp or constant-rate-of-change input. This is commonly referred to as a constant-velocity input, even though the units may be °C/s as well as m/s. Although this type of input produces a transient response, it is mainly used to provide information about the system's steady-state performance and in particular about steady-state errors.

c) Sinusoidal or harmonic input. This provides information about the steady-state frequency response of a system.

15.2 Equivalent systems

Measuring and control systems contain many types of mechanical, electrical, hydraulic, and pneumatic devices interconnected in various ways. For purposes of analysis it is convenient to represent them by equivalent systems made up of standard elements such as masses, springs, dampers, resistances, capacitances etc. which are more easily analysed to give their dynamic equations.

   For example, most systems contain mass, and if the mass is moving in a linear manner it is convenient to consider a linear representation – i.e. all mass lumped together in one place; all stiffnesses such as flexure in girders, supports, and connections considered as a single equivalent spring; and all damping considered as viscous damping in a damper or dashpot. We then have an equivalent mass–spring–damper system.

   If the mass is rotating it is more convenient to consider inertia, torsional stiffness, and rotary dampers.

In process flow systems it is convenient to represent all capacitances as tanks and all resistance effects lumped together as valves.

Similarly, complex circuits containing many resistances may be represented by a single equivalent resistance for convenience of analysis.

15.3 First-order system responses

15.3.1 First-order systems

If a feedback link and a comparison element such as a differential lever are added to the servo-valve/cylinder arrangement considered in example 14.6 in the previous chapter, a closed-loop system results with a block diagram illustrated in fig. 15.1, where

θ_i = input motion

ϵ = error signal

θ_o = output motion

K_v = servo-valve characteristic

A = effective area of cylinder

and $D \equiv \dfrac{d}{dt}$

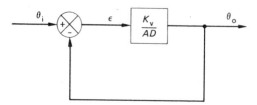

Fig. 15.1 Block diagram for a first-order hydraulic system

Using the basic closed-loop relationship from equation 13.1:

$$\frac{\theta_o}{\theta_i} = \frac{G}{1 + GH}$$

$$\text{closed-loop transfer operator} = \frac{\theta_o}{\theta_i} = \frac{K_v/AD}{1 + K_v/AD}$$

$$= \frac{1}{1 + (A/K_v)D}$$

$$\therefore \quad \frac{\theta_o}{\theta_i} = \frac{K}{1 + \tau D}$$

where $K = 1$ = steady-state gain or sensitivity

and $\tau = A/K_v$ = time constant

280

This system is therefore a *first-order system*, as defined in chapter 2.

Many other first-order systems and elements arise in measurement and control, e.g. thermocouples, resistance–capacitance electrical networks, level control, heat transfer through tank walls, etc. No matter what the system is, if its transfer operator can be arranged into the form

$$\frac{\theta_o}{\theta_i} = \frac{K}{1 + \tau D}$$

it will have a characteristic first-order type of response to any form of input.

Example 15.1 A pressure-measuring device consists of a variable restrictor (needle valve) in series with a bellows as shown in fig. 15.2. Assuming that the flow through the restrictor is directly proportional to the pressure drop across it, and that the change in capacitance of the bellows is negligible, show that the device has a first-order transfer operator.

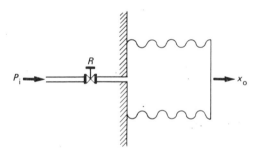

Fig. 15.2 A restrictor–bellows arrangement

Let p_i = input pressure (N/m^2)

R = valve characteristic (N s/m^5)

C = bellows capacitance (m^5/N)

p_b = pressure in bellows (N/m^2)

λ = bellows stiffness (N/m)

A = effective c.s.a. of bellows (m^2)

x_o = output displacement (m)

$\dot{v}$ = flow rate through the restrictor (m^3/s)

For the restrictor:

Flow rate through the restrictor is directly proportional to the pressure drop across it and inversely proportional to the restriction, i.e. $\quad \dot{v} = \dfrac{1}{R}(p_i - p_b)$ 15.1

For the bellows:

$\dfrac{\text{flow rate}}{\text{into the bellows}}$ = capacitance × rate of change of pressure

$\left[\text{electrical analogy:} \quad i = C\dfrac{dv}{dt} \right]$

i.e. $\quad \dot{v} = C\dfrac{dp_b}{dt}$

and, using the D operator,

$\quad \dot{v} = CDp_b$ 15.2

Equating equations 15.1 and 15.2,

$\dfrac{1}{R}(p_i - p_b) = CDp_b$

$\therefore \qquad p_i - p_b = RCDp_b$

$\qquad\qquad p_i = (1 + RCD)p_b$

$\qquad\qquad \dfrac{p_b}{p_i} = \dfrac{1}{(1 + RCD)}$ 15.3

For equilibrium in the bellows,

force due to pressure = spring force

$\qquad\qquad p_b \times A = \lambda \times x_o$

$\therefore \qquad\qquad \dfrac{x_o}{p_b} = \dfrac{A}{\lambda}$ 15.4

Combining equations 15.3 and 15.4,

$\dfrac{x_o}{p_i} = \dfrac{A/\lambda}{1 + RCD}$

In standard form, the transfer operator is

$\dfrac{x_o}{p_i} = \dfrac{K}{1 + \tau D}$

where   steady-state gain $K = A/\lambda$ (m³/N)

and   time constant $\tau = RC$ (s)

15.3.2 First-order response to a step input

It can be shown (see appendix A) that all systems with a first-order transfer operator have a characteristic exponential rise as illustrated in

282

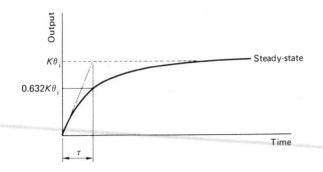

Fig. 15.3 First-order-lag step response

fig. 15.3 when subjected to a step input. Note that the magnitude of the output will be the same as the input only when the steady-state gain (or static sensitivity) K is unity.

Example 15.2   A thermocouple has a time constant of 3 s. Assuming that it has a first-order transfer operator, estimate the time for the thermocouple output to reach 98% of its final steady-state value when it is subjected to a step change in temperature.

Referring to fig. 2.7(b), a useful appoximation for first-order systems is that they take a time of 4 × (time constant) to get within 2% of the final steady-state value; therefore, in this case,

time to reach 98% = 4 × 3 s
                   = 12 s

15.3.3 *First-order response to a ramp input*
It can be shown (see appendix A) that all systems with a first-order transfer operator have the response illustrated in fig. 15.4 to a ramp input of the form $\theta_i = \Omega t$, where Ω is a constant.

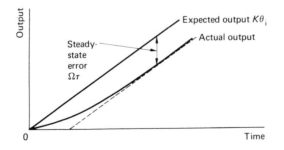

Fig. 15.4 First-order-lag ramp response

Note that

a) in the steady-state the actual output is always lagging behind the expected output by a constant amount known as the *steady-state error*, or as the *velocity lag* for a positional system;

b) this steady-state error is directly proportional to the time constant τ, so the larger τ the larger the magnitude of the error.

15.3.4 *First-order response to a sinusoidal input* (i.e. harmonic or frequency response)

In frequency response we are interested in two quantities:

i) the size of the output amplitude relative to the input amplitude, i.e. the amplitude ratio or modulus, denoted by $\left| \dfrac{\theta_o}{\theta_i} \right|$;

ii) the phase angle of the output relative to the input, denoted by ϕ.

It can be shown (see appendix A) that, for a first-order system,

$$\text{amplitude ratio } \left| \frac{\theta_o}{\theta_i} \right| = \frac{K}{\sqrt{[1 + (\omega\tau)^2]}} \qquad 15.5$$

and   phase angle $\phi = -\arctan \omega\tau$ \qquad\qquad 15.6

where the negative sign implies lagging phase and ω is the input frequency in rad/s.

These characteristics are illustrated in fig. 15.5.

Note that

a) the variable parameter is frequency in rad/s;

b) the amplitude ratio reduces as the frequency increases, i.e. a first-order lag element always attenuates the signal;

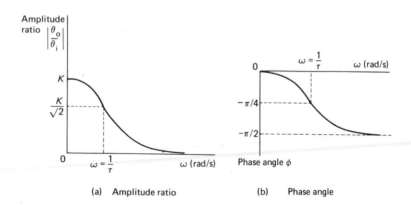

(a)   Amplitude ratio

(b)   Phase angle

Fig. 15.5 First-order-lag frequency response

c) the phase lag is $\pi/4$ at a frequency $\omega = 1/\tau$. This gives a method of obtaining the time constant of the system from experimental frequency-response data.

Example 15.3 A first-order measuring system with a steady-state gain of 1 is required to measure frequencies up to 150 Hz with an amplitude error of less than 10%. What is the maximum allowable time constant for the instrument?

Required amplitude ratio $\left|\dfrac{\theta_o}{\theta_i}\right| = 0.9$

Maximum $\omega = 2\pi \times 150 \,\text{rad/s}$

Therefore, using equation 15.5,

$$0.9 = \frac{1}{\sqrt{[1 + (300\pi\tau)^2]}}$$

$$1 + (300\pi\tau)^2 = \left(\frac{1}{0.9}\right)^2 = 1.23$$

$$\tau = \frac{\sqrt{(1.23 - 1)}}{300\pi \,\text{rad/s}} = 0.00051 \,\text{s}$$

∴ maximum allowable time constant $= 0.51 \,\text{ms}$

15.4 Second-order system responses

15.4.1 Second-order systems

a) Linear displacement systems
Many measuring systems and control devices can be represented by an equivalent system consisting of a spring, mass, and damper as shown in fig. 15.6, where

θ_i = input displacement (m)

θ_o = output displacement (m)

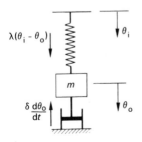

Fig. 15.6 A spring–mass–damper system

λ = spring stiffness (N/m)

δ = viscous damping coefficient (Ns/m)

and m = mass (kg)

The forces acting on the mass will be as follows.

Considering the spring; since both ends are free to move

$$\text{spring force} = \frac{\text{spring}}{\text{stiffness}} \times \frac{\text{displacement of one end of spring}}{\text{relative to other}}$$

$$= \lambda(\theta_i - \theta_o) \text{ downward}$$

Considering the dashpot: since one end is fixed, there will be a reaction force in an upward direction:

$$\text{damping force} = \text{damping coefficient} \times \text{velocity}$$

$$= \delta \frac{d\theta_o}{dt}$$

Using Newton's second law,

$$\Sigma \text{ forces} = \text{mass} \times \text{acceleration}$$

i.e. $$\Sigma F = ma$$

and $$a = \frac{d^2\theta_o}{dt^2}$$

$$\lambda(\theta_i - \theta_o) - \delta \frac{d\theta_o}{dt} = m \frac{d^2\theta_o}{dt^2}$$

Therefore, in terms of the D operator,

$$\lambda(\theta_i - \theta_o) - \delta D\theta_o = mD^2\theta_o$$

$$\lambda\theta_i = mD^2\theta_o + \delta D\theta_o + \lambda\theta_o$$

Dividing through by λ, we get

$$\theta_i = \left(\frac{m}{\lambda}D^2 + \frac{\delta}{\lambda}D + 1\right)\theta_o$$

and the transfer operator is

$$\frac{\theta_o}{\theta_i} = \frac{1}{(m/\lambda)D^2 + (\delta/\lambda)D + 1} \tag{15.7}$$

The standard second-order form, introduced in chapter 2, is

$$\frac{\theta_o}{\theta_i} = \frac{1}{(1/\omega_n^2)D^2 + (2\zeta/\omega_n)D + 1} \tag{15.8}$$

Comparing equations 15.7 and 15.8,

$$\text{undamped natural frequency } \omega_n = \sqrt{\frac{\lambda}{m}} \qquad\qquad 15.9$$

and

$$\frac{2\zeta}{\omega_n} = \frac{\delta}{\lambda}$$

∴ $$\text{damping ratio } \zeta = \frac{\delta\omega_n}{2\lambda} = \frac{\delta}{2\lambda}\sqrt{\frac{\lambda}{m}} \qquad\qquad 15.10$$

$$= \frac{\delta}{2\sqrt{(\lambda m)}}$$

Note that

a) the undamped natural frequency of a system can be increased by increasing the stiffness λ or by reducing the mass m, and vice versa;
b) the quantity $2\sqrt{(\lambda m)}$ in the denominator of equation 15.10 is equal to the critical damping coefficient (δ_c) of the system, i.e. the amount of damping required for there to be just no overshoot when the system is subjected to a step input.

Example 15.4   If the pen arrangement for a recorder has a mass of 5 g, determine the percentage reduction in pen mass to increase the natural frequency of the recorder by 20%. Assume that the pen motion can be considered as a second-order system.

From equation 15.9 we have

$$\omega_n = \sqrt{\frac{\lambda}{m}}$$

Using subscript 1 for initial and 2 for final values,

$$\omega_{n_1}^2 = \frac{\lambda}{m_1} \qquad\qquad 15.11$$

and $$\omega_{n_2}^2 = \frac{\lambda}{m_2} \qquad\qquad 15.12$$

Dividing equation 15.11 by equation 15.12,

$$\frac{\omega_{n_1}^2}{\omega_{n_2}^2} = \frac{m_2}{m_1}$$

∴ $$m_2 = m_1 \times \left[\frac{\omega_{n_1}}{\omega_{n_2}}\right]^2$$

But required $\omega_{n_2} = 1.2\omega_{n_1}$

$$\therefore \quad m_2 = m_1 \times \left[\frac{\omega_{n_1}}{1.2\omega_{n_1}} \right]^2$$

$$= \frac{m_1}{1.44} = 0.694m_1$$

$$\therefore \quad \% \text{ reduction in mass} = \frac{m_1 - 0.694m_1}{m_1} \times 100\%$$

$$= 30.6\%$$

b) Rotational systems
The rotational system analogous to the linear displacement system shown in fig. 15.6 is illustrated in fig. 15.7.

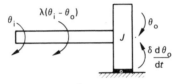

Fig. 15.7 A rotational system

By a similar analysis to that for the linear system, it can be shown that

$$\text{undamped natural frequency } \omega_n = \sqrt{\frac{\lambda}{J}} \text{ rad/s}$$

where λ = torsional stiffness (Nm/rad)
and J = system inertia (kgm²)

c) Remote position-control system
The d.c. remote position-control system discussed in section 13.4.1 has the block diagram shown in fig. 15.8, where

K_p = potentiometer sensitivity (V/rad)

G = amplifier gain (V/V)

K_m = motor constant (Nm/V)

J_e = equivalent inertia (kgm²)

δ_e = equivalent viscous friction (Nms/rad)

and n = gear ratio

288

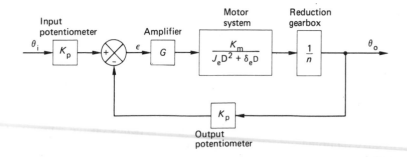

Fig. 15.8 Block diagram for a d.c. servo-system

Using the basic closed-loop relationship

$$\frac{\theta_o}{\theta_i} = \frac{G}{1 + GH} \quad \text{from equation 13.1}$$

the closed-loop transfer operator for the system is given by

$$\frac{\theta_o}{\theta_i} = \frac{\dfrac{K_p G K_m}{n(J_e D^2 + \delta_e D)}}{1 + \dfrac{K_p G K_m}{n(J_e D^2 + \delta_e D)}}$$

$$= \frac{K_p G K_m / n}{J_e D^2 + \delta_e D + K_p G K_m / n}$$

$K_p G K_m / n$ is referred to as the open-loop gain of the system. Putting this equal to K, we get

$$\frac{\theta_o}{\theta_i} = \frac{K}{J_e D^2 + \delta_e D + K}$$

Dividing through by K,

$$\frac{\theta_o}{\theta_i} = \frac{1}{(J_e / K)D^2 + (\delta_e / K)D + 1} \qquad 15.13$$

Comparing equation 15.13 with the standard second-order form of equation 15.8,

undamped natural frequency $\omega_n = \sqrt{\dfrac{K}{J_e}}$

and

$$\frac{2\zeta}{\omega_n} = \frac{\delta_e}{K}$$

Note that

a) the open-loop gain K is equivalent to the torsional stiffness of the rotational system discussed above;
b) the undamped natural frequency and hence the speed of response of the system can be changed by altering the open-loop gain K.

15.4.2 Second-order response to a step input

The response of all second-order systems depends on the amount of effective damping present. As explained in chapter 2, the usual method employed is to consider a damping ratio ζ, where

$$\text{damping ratio } \zeta = \frac{\text{damping coefficient for system}}{\text{critical damping coefficient for system}}$$

Critical damping is the amount of damping which just results in no overshoot when the system is subjected to a step input.

$\zeta = 1$ is therefore the *critically damped* case;

$\zeta > 1$ is the *heavy* or *overdamped* case, when no overshoot or oscillations occur; and

$\zeta < 1$ is the *light* or *underdamped* case, when overshoot and oscillation will occur.

Typical second-order responses to a step input for various values of damping ratio are shown in fig. 2.11.

Note that

a) the system responds faster (i.e. response time decreases) as the damping ratio is reduced, but overshoot and settling time are increased;
b) the magnitude of the overshoot is related to the damping ratio as illustrated in fig. 2.12.

The periodic time and frequency of damped oscillations are important parameters which will be illustrated in the following example.

Example 15.5 An internal-combustion engine fitted with a speed-regulating control system gave the response curve shown in fig. 15.9 when subjected to a step input. From the response curve, determine (a) the system rise time, (b) the 5% settling time for the system, (c) the percentage overshoot of the system, (d) the system damping ratio, (e) the periodic time of oscillation for the system, and (f) the damped frequency of the system, in hertz.

Referring to fig. 15.9:

a) Rise time is the time to get from 10% to 90% of the final steady-state value and is 0.6 s.
b) Time to get within 5% of final steady-state value = 11.4 s.

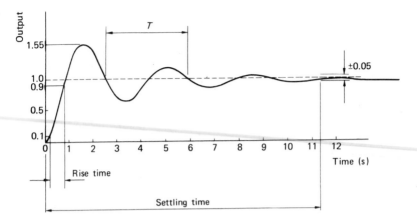

Fig. 15.9 The response curve for example 15.5

c) Percentage overshoot $= [(1.55 - 1)/1] \times 100\%$

$$= 55\%$$

d) Using fig. 2.12, for 55% overshoot $\zeta = 0.18$.

e) The time for one complete oscillation, i.e. the periodic time, $T = 3.4\,\text{s}$.

f) Damped frequency $f = 1/T = 0.29\,\text{Hz}$.

15.4.3 Second-order response to a ramp input
Typical responses to a ramp input are shown in fig. 15.10.
  Note that

a) the responses again depend on the magnitude of the damping ratio;
b) as with first-order systems, the steady-state output is lagging behind
   the input – i.e. there is a steady-state error;
c) the steady-state error is directly proportional to the damping ratio.

15.4.4 Second-order frequency response
As shown in appendix B, the frequency response for a second-order
system expressed in standard form is given by

$$\text{amplitude ratio} \quad \frac{\theta_o}{\theta_i} = \frac{1}{\sqrt{\{[1 - (\omega/\omega_n)^2]^2 + (2\zeta\omega/\omega_n)^2\}}} \qquad 15.14$$

$$\text{and} \quad \text{phase angle } \phi = -\arctan \frac{2\zeta\omega/\omega_n}{1 - (\omega/\omega_n)^2} \qquad 15.15$$

where ω is the input frequency in rad/s and the other symbols are as
defined earlier. These characteristics are illustrated in fig. 2.13.

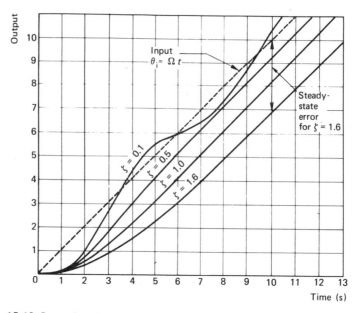

Fig. 15.10 Second-order ramp responses

Note that

a) the responses depend on the damping ratio;
b) for values of $\zeta < 0.7$ resonance occurs – i.e. the amplitude ratio is greater than 1, which means that the output amplitude is greater than the input;
c) the resonant frequency, i.e. the frequency at which maximum amplitude ratio occurs, is less than the natural frequency except for $\zeta = 0$;
d) from equation 15.15, it can be seen that the maximum phase lag is $180°$, and a phase lag of $90°$ occurs at an input frequency equal to the natural frequency, i.e. $\omega/\omega_n = 1$.

Example 15.6 It is usually arranged that a u.v. recording galvanometer has a damping ratio $\zeta = 0.64$. With this value of damping, determine the percentage amplitude error in the recorder output for an input signal varying at frequencies of (a) $0.6\omega_n$ and (b) $0.9\omega_n$.

a) Using equation 15.14,

$$\left| \frac{\theta_o}{\theta_i} \right| = \frac{1}{\sqrt{\{[1 - (0.6\omega_n/\omega_n)^2]^2 + [2 \times 0.64 \times 0.6\omega_n/\omega_n]^2\}}}$$

292

$$= \frac{1}{\sqrt{\{[1 - 0.36]^2 + 0.768^2\}}}$$

$$= 1.0003$$

Percentage error $= \dfrac{(1.0003 - 1)}{1} \times 100\%$

$$= 0.03\% \text{ high}$$

b) Again using equation 15.14 gives

$$\left| \frac{\theta_o}{\theta_i} \right| = 0.856$$

Percentage error $= \dfrac{(1 - 0.856)}{1} \times 100\%$

$$= 14.4\% \text{ low}$$

i.e. the 'flat' frequency response is up to 60% of the natural frequency of the galvanometer.

15.5 Control-system responses

Simple control systems may be represented by first- or second-order transfer operators and hence will produce the standard responses already outlined. More complex control systems will have higher-order transfer operators but the step, ramp, and harmonic responses tend to be of a similar form to those of second-order systems.

Although the system response is influenced by the amount of damping present, the most important parameter governing the response is the *open-loop gain* of the system. Facilities are available within the control system, usually at the controller, for manually adjusting the value of this open-loop gain. A high gain on the controller means that for a certain magnitude of error signal there will be a large controlling signal generated, while a low gain means that for the same error only a small controlling signal will be produced.

Increasing the gain has the same effect as reducing the damping ratio on a second-order system, i.e. the response time is reduced but at the expense of increased overshoot and longer settling time. A further problem with higher-order systems is that, if the gain is increased too far, instability can be produced – i.e. a constantly increasing output.

Reducing the gain has the same effect as increasing the damping ratio.

Example 15.7 If the open-loop gain of the speed-regulating control system of example 15.5 was (a) doubled then (b) halved, sketch possible response curves for the system.

a) The system would have a quicker response, larger overshoot, and more oscillation, as illustrated in fig. 15.11.

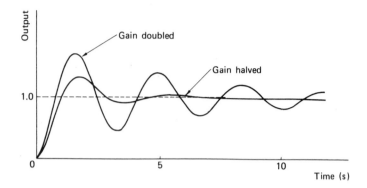

Fig. 15.11 Response curves for example 15.7

b) The system would have a slower response, smaller overshoot, and less oscillation, as illustrated in fig. 15.11.

Exercises on chapter 15

1 A thermometer has a thermal time constant of 5s. Determine (a) the approximate time it will take to get within 2% of the temperature of a fluid into which it is placed and (b) the steady-state error to be expected if the temperature of the fluid is increasing at 8°C per min.   [20s; 0.67°C]

2 Explain what is meant by (a) a first-order system and (b) a second-order system. Sketch the type of response expected when these systems are subjected to step and ramp inputs.

3 Define the following terms: (a) time constant, (b) steady-state error, (c) critically damped system, (d) underdamped system, and (e) periodic time.

4 In a frequency-response test on a first-order system, a phase lag of 45° is obtained at a frequency of 2Hz. What is the system time constant?   [0.08s]

5 The specification for a temperature-control system requires that the thermocouple employed must get within a tolerance band of 1°C, for a sudden change in temperature of 50°C, within 8s. What is the maximum permissible time constant for the thermocouple?   [2s]

6 A second-order system is described by the differential equation

$$0.1 \frac{d^2y}{dt^2} + 0.2 \frac{dy}{dt} + 2.5y = 2.5x$$

where x is the input of the system and y the output.

By comparing the equation with the standard second-order form, determine (a) the system undamped natural frequency (in Hz) and (b) the damping ratio.   [0.796Hz; 0.2]

7 An ultra-violet recorder uses moving-coil galvanometers. These can be represented by rotors acting against torsional springs and damped by viscous fluid, with the following differential equation of motion:

$$J \frac{d^2\theta_o}{dt^2} + \delta \frac{d\theta_o}{dt} + \lambda\theta_o = \lambda\theta_i$$

If the inertia J is $1\,\text{mg}\,\text{m}^2$ and the torsional stiffness λ is $9.87\,\text{Nm/rad}$, determine:

a) the natural undamped frequency of the galvanometer,
b) the viscous damping coefficient δ required to critically damp the movement, and
c) the value of the viscous damping coefficient δ to give a damping ratio of 0.64.

Sketch the galvanometer responses to step inputs for (b) and (c).   [$500\,\text{Hz}$; $6.3\,\text{mN}\,\text{ms/rad}$; $4.03\,\text{mN}\,\text{ms/rad}$]

8 The d.c. remote position-control system shown in fig. 15.8 has the following values for the parameters:

Potentiometer sensitivity $50\,\text{V/rad}$
Amplifier gain $8.57\,\text{mA/V}$
Motor constant $0.7\,\text{Nm/A}$
Equivalent inertia $0.3\,\text{gm}^2$
Equivalent viscous friction $3\,\text{mN}\,\text{ms/rad}$
Reduction gear ratio 10

Determine (a) the system undamped natural frequency in hertz, and (b) the effective damping ratio of the system.   [$1.59\,\text{Hz}$; 0.5]

Appendices

The solutions given are practical but not mathematically rigorous.

Appendix A: response of a first-order lag

Consider a typical first-order lag element with a transfer operator given by

$$\frac{\theta_o}{\theta_i} = \frac{K}{1 + \tau D}$$

In differential-equation form, this is

$$(1 + \tau D)\theta_o = K\theta_i$$

a) Transient response (complementary function)

This is given by

$$(1 + \tau D)\theta_{o,t} = 0$$

Assuming a solution of the form $\theta_{o,t} = Ae^{mt}$,

$$(1 + \tau D)Ae^{mt} = 0$$

$$Ae^{mt} + m\tau Ae^{mt} = 0$$

$$\therefore \quad Ae^{mt}(1 + m\tau) = 0$$

and the auxiliary equation is given by

$$(1 + m\tau) = 0$$

$$\therefore \qquad m = -1/\tau$$

and $\quad \theta_{o,t} = Ae^{-t/\tau}$ \hfill A.1

b) Steady-state response (particular integral)

This is given by

$$(1 + \tau D)\theta_{o,ss} = K\theta_i$$

$$\therefore \qquad \theta_{o,ss} = (1 + \tau D)^{-1}K\theta_i$$

Using the binomial expansion

$$(1 + x)^n = 1 + \frac{nx}{1!} + \frac{n(n - 1)x^2}{2!} + \dots$$

this gives
$$\theta_{o,ss} = (1 - \tau D + \text{terms in } D^2 \text{ and higher}) K\theta_i$$

i) Step input
If the input θ_i is a step of constant magnitude θ_i,
$$\theta_{o,ss} = (1 - \tau D + \text{terms in } D^2 \text{ and higher}) K\theta_i$$
$$\therefore \quad \theta_{o,ss} = K\theta_i \quad (\text{since } D\theta_i = 0)$$

The complete response is given by the sum of the transient response and the steady-state response;

i.e. $\theta_o = \theta_{o,t} + \theta_{o,ss}$

$\therefore \qquad \theta_o = Ae^{-t/\tau} + K\theta_i$

Applying initial conditions $\theta_o = 0$ at time $t = 0$,

$0 = A + K\theta_i$

$A = -K\theta_i$

$\therefore \quad \theta_o = K\theta_i(1 - e^{-t/\tau})$

ii) Ramp input
If the input θ_i is a ramp of the form $\theta_i = \Omega t$, where Ω is a constant,
$$\theta_{o,ss} = (1 - \tau D + \text{terms in } D^2 \text{ and higher}) K\Omega t$$
$$\therefore \quad \theta_{o,ss} = K\Omega t - K\Omega\tau \quad (\text{since } D^2\Omega t = 0)$$

The complete response is given by
$$\theta_o = \theta_{o,t} + \theta_{o,ss}$$

i.e. $\theta_o = Ae^{-t/\tau} + K\Omega t - K\Omega\tau$

Applying initial conditions $\theta_o = 0$ at time $t = 0$,

$0 = A - K\Omega\tau$

$A = K\Omega\tau$

$\therefore \qquad \theta_o = K\Omega t - K\Omega\tau (1 - e^{-t/\tau})$

iii) Harmonic input
The transient response for a harmonic input is the same as equation A.1; however, in frequency-response testing we are interested only in the steady-state response.

It can be shown that the steady-state frequency response of any system can be obtained by replacing D in the transfer operator by $j\omega$, where ω is the input frequency in rad/s and $j = \sqrt{(-1)}$.

Therefore, for a first-order system the frequency response is given by

$$\frac{\theta_o}{\theta_i} = \frac{K}{1 + j\omega\tau} \qquad\qquad \text{A.2}$$

In frequency response we are interested in the amplitude ratio and phase relationship between the output and input. These are given by the modulus and argument respectively of equation A.2.

Equation A.2 can be converted from the form $a + jb$ to the form $R\underline{/\theta}$ by using the relationships

$$\text{modulus } R = \sqrt{(a^2 + b^2)}$$

and   argument $\theta = \arctan(b/a)$

Therefore, in this case,

$$\text{amplitude ratio} \left| \frac{\theta_o}{\theta_i} \right| = \frac{\sqrt{(K^2 + 0^2)}}{\sqrt{[1^2 + (\omega\tau)^2]}}$$

i.e.

$$\left| \frac{\theta_o}{\theta_i} \right| = \frac{K}{\sqrt{[1 + (\omega\tau)^2]}}$$

and   phase angle $\phi = \arctan \dfrac{0}{K} - \arctan \dfrac{\omega\tau}{1}$

[note: the minus sign arises because the term $(1 + j\omega\tau)$ is in the denominator of equation A.2]

i.e. $\phi = -\arctan \omega\tau$ (lagging)

Appendix B: response of a second-order (quadratic) lag
Consider a typical second-order lag element with a transfer operator given by

$$\frac{\theta_o}{\theta_i} = \frac{K}{(1/\omega_n^2)D^2 + (2\zeta/\omega_n)D + 1}$$

In differential-equation form, this is

$$\left(\frac{1}{\omega_n^2} D^2 + \frac{2\zeta}{\omega_n} D + 1 \right) \theta_o = K\theta_i$$

a) Transient response
This is given by

$$\left(\frac{1}{\omega_n^2} D^2 + \frac{2\zeta}{\omega_n} D + 1 \right) \theta_{o.t} = 0$$

Assuming a solution of the form $\theta_{o.t} = Ae^{mt}$

$$\left(\frac{1}{\omega_n^2} D^2 + \frac{2\zeta}{\omega_n} D + 1 \right) Ae^{mt} = 0$$

$$\therefore \quad Ae^{mt} \left(\frac{1}{\omega_n^2} m^2 + \frac{2\zeta}{\omega_n} m + 1 \right) = 0$$

and the auxiliary equation is given by

$$\left(\frac{1}{\omega_n^2}m^2 + \frac{2\zeta}{\omega_n}m + 1\right) = 0$$

$$\therefore \quad m = \frac{-2\zeta/\omega_n \pm \sqrt{[(2\zeta/\omega_n)^2 - 4/\omega_n^2]}}{2/\omega_n^2}$$

$$= -\zeta\omega_n \pm \sqrt{[\zeta^2\omega_n^2 - \omega_n^2]}$$

$$= -\zeta\omega_n \pm \sqrt{[\omega_n^2(\zeta^2 - 1)]} \qquad \text{B.1}$$

There are three possible solutions, depending on the magnitude of the damping ratio ζ.

i) Heavy damping, i.e. $\zeta > 1$

$$\theta_{o.t} = Ae^{-\zeta\omega_n t - [\omega_n\sqrt{(\zeta^2-1)}]t} + Be^{-\zeta\omega_n t + [\omega_n\sqrt{(\zeta^2-1)}]t}$$

ii) Critical damping, i.e. $\zeta = 1$
Equation B.1 becomes

$$m = -\zeta\omega_n$$

$$\therefore \quad \theta_{o.t} = (At + B)e^{-\zeta\omega_n t}$$

iii) Light damping, i.e. $\zeta < 1$
Equation B.1 becomes

$$m = -\zeta\omega_n \pm j\omega_n\sqrt{(1 - \zeta^2)}$$

and $\quad \theta_{o.t} = Ae^{-\zeta\omega_n t - j[\omega_n\sqrt{(1-\zeta^2)}]t} + Be^{-\zeta\omega_n t + j[\omega_n\sqrt{(1-\zeta^2)}]t}$

Replacing the complex exponentials by sines and cosines, this becomes

$$\theta_{o.t} = e^{-\zeta\omega_n t}\{A\cos[\omega_n\sqrt{(1 - \zeta^2)}]t + B\sin[\omega_n\sqrt{(1 - \zeta^2)}]t\} \qquad \text{B.2}$$

b) Steady-state response
This is given by

$$\left(\frac{1}{\omega_n^2}D^2 + \frac{2\zeta}{\omega_n}D + 1\right)\theta_{o.ss} = K\theta_i \qquad \text{B.3}$$

i) Step input
If the input θ_i is a step of constant magnitude θ_i then, in the steady-state, D terms and higher are zero.

$$\therefore \quad \theta_{o.ss} = K\theta_i$$

The complete response is given by

$$\theta_o = \theta_{o.t} + \theta_{o.ss}$$

Considering only the lightly damped case,

$$\theta_o = K\theta_i + e^{-\zeta\omega_n t}\{A\cos[\omega_n\sqrt{(1-\zeta^2)}]t + B\sin[\omega_n\sqrt{(1-\zeta^2)}]t\}$$

Applying initial conditions $\theta_o = 0$ and $d\theta_o/dt = 0$ at time $t = 0$ gives

$$A = -K\theta_i \quad \text{and} \quad B = -\frac{K\theta_i\zeta}{\sqrt{(1-\zeta^2)}}$$

$$\therefore \quad \theta_o = K\theta_i\left[1 - e^{-\zeta\omega_n t}\{\cos[\omega_n\sqrt{(1-\zeta^2)}]t + \frac{\zeta}{\sqrt{(1-\zeta^2)}}\sin[\omega_n\sqrt{(1-\zeta^2)}]t\}\right]$$

The quantity $\omega_n\sqrt{(1-\zeta^2)}$ gives the damped natural frequency of the system, ω_{nd}. This is the frequency (in rad/s) at which the *damped* system would oscillate when disturbed.

ii) Ramp input

If the input θ_i is a ramp of the form $\theta_i = \Omega t$, where Ω is a constant, in the *steady-state* D² terms and higher are zero and equation B.3 becomes

$$\left(1 + \frac{2\zeta}{\omega_n}D\right)\theta_{o.ss} = K\Omega t$$

$$\theta_{o.ss} = \left(1 + \frac{2\zeta}{\omega_n}D\right)^{-1}K\Omega t$$

Expanding by the binomial theorem, this becomes

$$\theta_{o.ss} = \left(1 - \frac{2\zeta}{\omega_n}D + \text{terms in D}^2 \text{ and higher}\right)K\Omega t$$

$$= K\Omega t - \frac{2K\Omega\zeta}{\omega_n}$$

The complete response is given by

$$\theta_o = \theta_{o.t} + \theta_{o.ss}$$

and applying the same initial conditions as in (i), for the lightly damped case,

$$\theta_o = K\left[\Omega t - \frac{2\zeta\Omega}{\omega_n} + \frac{2\zeta\Omega e^{-\zeta\omega_n t}}{\omega_n}\left\{\cos[\omega_n\sqrt{(1-\zeta^2)}]t + \frac{2\zeta^2 - 1}{2\zeta\sqrt{(1-\zeta^2)}}\sin[\omega_n\sqrt{(1-\zeta^2)}]t\right\}\right]$$

iii) Harmonic input

Again we are interested only in the steady-state response, and this is given by replacing D by $j\omega$ as before.

$$\therefore \quad \frac{\theta_0}{\theta_i} = \frac{K}{(1/\omega_n^2)(j\omega)^2 + (2\zeta/\omega_n)(j\omega) + 1}$$

$$= \frac{K}{[1 - (\omega/\omega_n)^2] + j(2\zeta\omega/\omega_n)}$$

$\therefore$ amplitude ratio
$$\left| \frac{\theta_0}{\theta_i} \right| = \frac{\sqrt{(K^2 + 0^2)}}{\sqrt{\{[1 - (\omega/\omega_n)^2]^2 + (2\zeta\omega/\omega_n)^2\}}}$$

i.e.
$$\left| \frac{\theta_0}{\theta_i} \right| = \frac{K}{\sqrt{\{[1 - (\omega/\omega_n)^2]^2 + 4\zeta^2(\omega/\omega_n)^2\}}}$$

and phase angle $\phi = \arctan \dfrac{0}{K} - \arctan \dfrac{2\zeta(\omega/\omega_n)}{1 - (\omega/\omega_n)^2}$

$$= -\arctan \frac{2\zeta(\omega/\omega_n)}{1 - (\omega/\omega_n)^2}$$

Appendix C: decibel (dB) notation

Two powers P_1 and P_2 are said to differ by N 'bels' when

$$\frac{P_2}{P_1} = 10^N$$

i.e. $N = \log_{10} \dfrac{P_2}{P_1}$ bels

In practice, the bel is found to be inconveniently large, and 0.1 of it – the decibel – is more often used; thus

$$N = 10\log_{10} \frac{P_2}{P_1} \, \text{dB}$$

The decibel notation thus provides a convenient method of expressing ratios of powers (and amplitudes) which may extend over an enormous range. It is used to express power and amplitude gain or loss in electronic amplifiers, control systems, acoustics, and vibration studies. For example, in acoustics the range of audible intensity (power per unit area) extends from the threshold of hearing to the threshold of pain (in the ear), an increase of 10^{14}. It is clearly difficult to express and use such large ratios using simple arithmetic. Using the dB notation allows the ratio between the above two thresholds to be rewritten as

$$N = 10 \log_{10} \frac{10^{14}}{1}$$

$= 140 \text{dB}$ which is in a much simpler form

Amplitude gain in dB

Consider an electronic amplifier in which two powers P_1 and P_2 may be compared by observing the amplitudes of the corresponding voltages developed across a given impedance. If the input and output impedances of the amplifier are *equal*, the power ratio will be proportional to the square of the voltage. Hence, if P_1 and P_2 are the input and output powers of the amplifier,

$$\text{power gain} = 10 \log_{10} \left(\frac{P_2}{P_1} \right) \text{dB}$$

$$= 10 \log_{10} \left(\frac{V_2}{V_1} \right)^2$$

$$\text{voltage amplitude gain} = 20 \log_{10} \frac{V_2}{V_1} \text{ dB} \qquad \text{C.1}$$

This relationship has, through popular usage, been used to express the voltage gain of amplifiers without considering the condition of equality between the input and output impedances. Despite this restriction, equation C.1 finds wide application in electronics and control engineering as a method of expressing amplitude ratio, although its use is not technically correct.

Further reading

General texts

Barney, G. C., *Intelligent Instrumentation* (Prentice Hall)

Doebelin, E. O., *Measurement Systems – Application and Design* (McGraw-Hill)

Harriott, P., *Process Control* (McGraw-Hill)

Healey, M., *Principles of Automatic Control* (Edward Arnold)

Holland, J. M., *Basic Robotic Concepts* (H. W. Sams & Co.)

Jones, B. E., *Instrumentation, Measurement and Feedback* (McGraw-Hill)

O'Higgins, P. J., *Basic Instrumentation* (McGraw-Hill)

Shinskey, F. G., *Process Control Systems* (McGraw-Hill)

Warring, R. H., *Handbook of Noise and Vibration Control* (Trade and Technical Press)

Index

summer amplifier, 61
swivelling jet, 268
symmetrical bridge, 147
synchros, 259
system responses, 279–95
  harmonic, 15, 21, 279, 291
  ramp, 279, 283, 291
  steady-state, 279, 296, 299
  step, 14, 19, 279, 282, 290
  transient, 279

tachogenerator
  a.c., 129
  d.c., 132
tachometer
  drag-cup, 131
  mechanical, 128
tape recorder, *see* magnetic-tape
  recorder
telemetry, 240
temperature, 221–41
  coefficient of resistance, 231
  control, 274
  gradient, 234
  triple-point, 221
thermal-conductivity gauge, 191
thermistors, 35, 233, 238
thermocouples, 224–30
  base-metal, 226
  cold junction, 228
  construction, 225
  extension leads, 229
  hot junction, 224
  ranges of, 227
  rare-metal, 227
thermocouple (vacuum) gauge, 192, 196
thermometers
  bimetallic, 76, 224
  gas, 224
  liquid-expansion, 222, 238
  liquid-in-glass, 222, 238
  thermocouples, 224–30
thermopile, 230
thermowell, 225
time constant, 14, 226, 283
timer/counter, 126
tolerance, 10
torr, 177
torsional stiffness, 288
transducers, 3, 27–50
  capacitive, 37, 112, 185
  classification, 27
  definition of, 27
  electromagnetic, 43, 207
  inductive, 39, 110, 115, 185
  linear variable-differential, 41, 108

mechanical, 46
photoelectric, 45
piezo-electric, 42, 206
pressure, 185
quartz, 188, 206
resistance, 29, 105, 186, 258
seismic-mass, 204
thermoelectric, 45
transformers, 41, 108
vibration, 203
transfer operator, 243, 269, 293
transient response, 278, 296, 298
translational mode, 199
triple point, 221
turbine–generator set, 244

ultra-violet galvanometers, 79–88
ultra-violet recorder, 79
unbonded strain gauge, 144
undamped natural frequency, 17, 289
underdamped system, 18, 290
underlapped valve characteristic, 267
unity feedback, 252
unmonitored system, 244
U-tube manometer, 180

valve actuator, 262
valve centre characteristic, 266
valve flow characteristic, 264
vapour-deposited gauge, 186
variable-reluctance transducer, 39
velocity lag, 284
velocity, reference, 200
velocity transducer, 44
vibration
  acceleration, 200
  amplitude, 201
  effects, 199
  level, 199
  measurement, 199–211
    reed, 204
    seismic-mass, 204, 117
  spectra, 202
  transducers, 203
viscous damping coefficient, 286
V-port plugs, 264

Ward–Leonard control system, 272
water-cooled transducer, 189
ways, valve, 266
weighbridge, 161
weighing, 160, 173
  system, 174
well manometer, 181